"Earthy, economical prose . . . wisdom and subtle understanding. . . Evolution, as *Mr. Darwin's Shooter* demonstrates, is driven by forces more nuanced and mysterious than the crude survival of the fittest."
—Walter Kirn, *Time*

"A richly imagined period piece about the conundrums of discovery . . . McDonald deftly explores the uneasy balance between science and belief."
—*Outside*

"McDonald brings Covington to life in such a vivid way that the ink on the page seems mixed with sea brine and blood. From the first page, there is no doubt that this character and his story are in the hands of a strong, thoughtful writer." —*The Hartford Courant*

"A book that breathes life, with something solid and heartbreaking at is core . . . Syms Covington may be a footnote in Darwin's text, but in McDonald's hands the spear carrier with the restless mind and troubled heart springs into splendid, uncompromising life."
— *The News & Observer* (Raleigh)

"McDonald has plucked [Covington] from near anonymity and given him flesh, blood, and best of all, a compelling voice."
—*Providence Sunday Journal*

"A spectacular tale of nineteenth-century exploration and the conflict between science and religion, all based on Charles Darwin's famous voyage of discovery . . . an impressively learned novel . . . The triumphant characterization of the embattled Syms—a man in whom intellect, appetite, and religious impulse coexist credibly and explosively—is deepened as the story plays increasingly intriguing variations on the concepts of singularity and class, both as scientific definition and as a more clumsy way of clarifying human relations. Brilliant work: an Australian novel that merits comparison with Patrick White's masterpiece, *A Fringe of Leaves*."
—*Kirkus Reviews* (starred review)

"A wild adventure that has something of the verve of Robert Louis Stevenson and the lyricism of Herman Melville . . . McDonald is a passionate writer who loves the tastes and textures of the world, but never loses touch with the shifting, tempestuous emotions of his characters."
—*Nashville Scene*

"An impressively researched book, rich in the tangled issues that surround Darwin and his work, especially its shock to Victorian religious sensibilities . . . McDonald constantly surprises. His prose is ebullient, even boisterous, grabbing the reader with language so vivid and original, alternately comic and tragic, that it reads like something out of Dickens. McDonald never falls into a dry historical tone, simply because he refuses to lose the sweaty, angry, sad, violent reality of life."        *—Book Page*

"Brilliant . . . a book that seems to be not only *about* the early nineteenth-century, but *of* it."        *—San Jose Mercury News*

"Without raising any suspicion of a moral agenda, Roger McDonald accomplishes one of the greatest moral possibilities of fiction—to imaginatively restore the lives that history has left out. . . . Covington is the ideal mirror of the loss Western culture suffered when it came against the theory of evolution, and realized the threat to its spiritual life. . . . I cannot compliment highly enough this subtle, moving, intelligent book. A fascinating choice of subject is developed in a lyrical, original way."
*—Rocky Mountain News*

"By reimagining this mysterious presence at the side of one of the most important scientists of all time, McDonald has created a dense, remarkably complex character and a life fascinatingly caught between faith and science."        *—Minneapolis City Pages*

"A meticulously imagined plot and vivid characters, and written in a lovely prose that blends nineteenth-century English vernacular, Australian dialect, and Melvillean rhythms . . . In its combination of picaresque sea adventure and metaphysical inquiry, *Mr. Darwin's Shooter* manages to be both an absorbing and a thoughtful book."        *—Boston Phoenix Literary Supplement*

# Mr Darwin's
# Shooter

Also by Roger McDonald

FICTION:

1915
*Slipstream*
*Rough Wallaby*
*Water Man*
*The Slap*

NON-FICTION:
*Shearers' Motel*

AS EDITOR:
*Gone Bush*

ROGER McDONALD

*M*R *D*ARWIN'S
SHOOTER

GROVE PRESS
*New York*

The writing of this work was assisted by an Earnback Fellowship from the Australia Council, the Federal Government's arts funding and advisory body.

The epigraph from Bashö is taken from *On Love and Barley: Haiku of Bashö,* translated and introduced by Lucien Stryk, published by Penguin Classics, 1995.

*Originally published by Random House Australia Pty Ltd*
*Published simultaneously in Canada*
*Printed in the United States of America*

Library of Congress Cataloging-in-Publication Data

McDonald, Roger, 1941–
    Mr. Darwin's shooter / Roger McDonald.
      p.   cm.
    ISBN-10: 0-8021-4356-3
    ISBN-13: 978-0-8021-4356-3
    1. Covington, Syms, 1813–1861—Fiction. 2. Darwin, Charles, 1809–1882—Friends and associates—Fiction. 3. Beagle Expedition (1831–1836)—Fiction.   I. Title.
PR9619.3.M2514M7    1999
823—dc21                      98-36819

Design by Gayna Murphy

Grove Press
an imprint of Grove/Atlantic, Inc.
841 Broadway
New York, NY 10003
Distributed by Publishers Group West
www.groveatlantic.com

08 09 10 11 12   10 9 8 7 6 5 4 3 2 1

*To Elinor, Anna and Stella*
*with love*
*and to Susie with your spirit*
*shining*

'*N*ow darkness falls,'
   quail chirps,
'what use hawk-eyes?'

BASHO

# CONTENTS

# PROLOGUE

## *On a Dish of Milk Well-Crumbed*

# 1828

The day was hot and dusty with scattered leaves of poplars lining a towpath. A boy went swimming in green canal water, rolling himself belly-over, gulping and thrashing in pleasure. He beat the slowly moving water with the flat of his hand and floating facedown blew noisy bubbles.

Syms Covington was naked as a bulb, white and hairless except for a slicked-down tuft of red curls across the dome of his conspicuous head. At twelve years of age he was sturdy as a man and soon would become one, stretching in his bones until he reached a height of just six feet, and getting a strength across his shoulders and in his arms like a house beam squared from timber. Yet when Covington floated on his back between corridors of puffy summer clouds he felt small as a flea, and imagined he looked down on the earth. It made a field of blue for him to hop around in. He laughed and squeaked, never minding how cold the water was, and went swimming any time of year to win wagers or for the joy of it alone. Other times he took bread and cheese in a sack and wandered the fields. On summer nights he slept with a stone for his pillow like Jacob, waking in the moonlight and hearing a badger grunt and watching a hare strip bark from a sapling. He fought his fears on such nights and saw them come to nothing in the early light.

When he reached the gates of the lock he could hear water trickling far below. It came from a dark door. There were times when Covington had swum below that door and thought of the weight of water above him. He knew the gates were held by iron bars, ratchets, cogs, and by oakwood planks. But all the same, what if the weight of water broke them? When he thought about that he saw himself on the surface of the water, shooting away like a leaf, and his illusion of floating in the sky vanished. Then he knew the feeling of being tested against eternal punishment and knowing he was loved.

Upstream Covington began his play again, heading back to where two bundles of clothes awaited him on the bank, one bloodstained and filthy, the second lot as clean as hard scrubbing and hanging in the sun could make them. The canal became a river at that place, with willows trailing their branches and a water rat making a spline of ripples. It was a place to be cleansed of stains, except the boy had been in the water a good half hour and his forearms were still sticky with blood and flecks of fatty meat. He grabbed a handful of clay and scrubbed himself. He started singing. While he stood there, balancing in the mire, a man got up from under a tree near the lock-keeper's cottage and walked along the towpath.

The man wore a soft sailor's cap with curly black hair poking from underneath, and a red waist-jacket leaving his ribs bare in the heat. He was past thirty years of age, short of stature, with a rounded black beard composed of tight corkscrew curls. His sunken eyes were feverish, his red lips parched, and when he swallowed a prominent Adam's apple travelled up and down. He carried his belongings in a sailorman's sack hung over his shoulder, and when he reached a narrow bridge that was barely more than a plank with a handrail, he shifted the sack to the other shoulder and walked the plank with an assertive and derisive gait, giving a few hard bounces along its length. From there he watched Covington amusing himself. Daubing and day-dreaming the boy sang 'Barley Mow' in a sweet soprano as clear as any girl's, and this was remarkable because the sailor, whose name was John Phipps, had been thinking the boy looked like a shaved pig, and in the purity of the outburst asked God's forgiveness for such thoughts and said a prayer for the impressment of souls.

All that Saturday afternoon Covington had helped his father and brothers, hauling horsemeat from a wagon sticky with flies and chopping it into portions on a market table. His Pa was a Bedford butcher wielding a long knife and bringing unwanted carthorses to their knees in a welter of blood and callous humour. After the markets the boy did the scrubbing-down with a stiff broom and a tub of soapy water. He had smiling narrow eyes, dusty blue in colour, high cheekbones and a wide generous mouth. His nose was aquiline, his nostrils slightly flared, and the bones of his forehead were like a shield. When asked why he laboured with no pay when he slaved all week, a clerk in a leather-merchant's house, he brushed curls from his forehead and gave a shrug:

'Say the broom makes a good sound hitting the bristles against the stone. Say I feel like I'm drumming and making music for 'un.'

Covington's brothers and Pa at the end of their Saturday labours sank pots of dark ale, giving themselves winking blades of foam up their cheekbones. They earned them, in the boy's honest estimation.

'My boy,' said his Pa in return admiration, 'is a true old-time Covington, the most willing soul that ever lived.'

The steamy-breathed old man had bristly eyebrows flying back over his forehead, and prominent front teeth showing yellow and flat when he drew his lips back. Standing in his blood-brown boots he rocked back and forth as if hammered to the ground and twanging slightly with the force of his opinions.

He liked to call his son over and hold his head back in a playful grip, trickling bitter ale into his mouth and down his chin. Their people, he liked to boast, were Bedford notables in the time of Oliver Cromwell and their line went back past 1199, when they owned half a virgate of land. 'Of all the children of my bowels, Simon is the one that God has chosen to better his self, and lucky for us and ours.' His brothers passed the boy the ale-pot in the same rough animal-play. After drinking it down companionably, and staggering around to make them laugh, Covington returned to his sweeping with a light head. The others stood in shabby doorways with their shirt collars open, their belts loosened and slippery leather laces dangling. They were ready to kick their boots off and go crawling in a corner when they were too drunk to stand. But it would be a good long while before they were felled. Something about the Covingtons recalled animals associated with primitive man. The barely domesticated. Those spirits to the end. Say bullocks with clear foreheads and curly scruffs of hair from the ancient cave paintings of Spain and France—they were found in their lifetime—or strong-necked ponies from the same smoky walls, ten or twenty or thirty thousand years ago, pale-eyed and bristly-maned in the dawn of the roping, the taming, and the hard use of innocence in the aims of civilisation. Covington would one day think so, anyway. They were dirty-fisted hard-working men given to their pipes, their ale, their loud opinions—likewise to their routines of sudden mayhem, sharp knives, rolled back horses' eyes and clattering hooves. Being horse-butchers they were lower-placed than those who dealt with finished hides. But Covington never felt shame and pity for them, for while they were mired in blood they remembered they had souls, and Covington was of them truly—except that if he was to spend his whole life around them he would never find what he wanted.

In the deepest part of himself he knew what that was, and it meant setting off on a journey. A story tingled his arms to the fingertips and shook his shanks down to his toes with anxiety and restlessness. It was the *Pilgrim's Progress* that belonged to their town and countryside, telling of a sally away from Bedford in a great undertaking. It was all about walking and peering and finding, coming out from behind trees and passing down narrow rocky paths into darkness and light. It was all a great test for goodness of heart. Obstacles were to be met, most horrendous, and there were dangers of falling into an abyss. Black rivers were to be crossed. Vain and foolish strangers were to be put to rights.

John Bunyan's book was devotional reading in the house in Mill Lane from the time they were small. It cleansed them just to think of it. In Bedford and the nearby countryside you would think the very air breathed was old John Bunyan's. The long, sky-wide quality of the light and the feel of the chalk and clay came from Bunyan's pages. Likewise the water meadows and the winter floods, they all squelched and trickled with his words. Bunyan was preached without cease in their chapel and his language, whether imbued with ale or milkmaid's curds, always had the taste of the countryside and its pleasures and pitfalls in it:

*The next was a dish of milk well-crumbed. But Gaius said, Let the boys have that, that they may grow thereby.*

None of them took it quite as Covington did, with the seriousness of a promise and a passion of loyalty in his bones. At the age he was he regretted how all the great wars had finished when he was too young to fight them. He wished that monster Napoleon who'd been imprisoned on St Helena Island had lived to old age and given him the chance to stand in arms against him, the way Christian in Bunyan's pages had stood against Apollyon, who had scales like a fish, wings like a dragon, feet like a bear, and out of whose belly came fire and smoke. Death was not to be feared in such a spirit except in failure, and if Covington succeeded it would be a greatness overcoming all. He would be a boy hero like those at Trafalgar and stand on a splintering deck risking his life with every thump of a gun. Or he would advance with a bayonet, impaling native heads. He would rise in worth and join with the chosen of England—although, as John Bunyan put it, 'not at the first, nor second, nor third, nor fourth, nor fifth, no nor at the sixth time neither'. Because if you had a strong pull in any path of life then obstacles came at you to greet you.

His earliest memories were not of his mother nurturing him, but of a man with golden curls. His name was Christian. He had rosy cheeks and wore a raspberry-red jacket with gold buttons. The light shone through him by day, while at night his colours went dead as mud. On Sunday mornings he flew soaring over a stile and simultaneously looked back over his shoulder and met Covington's gaze with the bottle-brown of a single eye. He made a beckoning gesture with a crooked finger: 'Follow me.' He was made from coloured glass in a window setting, but the boy didn't know that, in his earliest conjecturing of the world, in which everything past the reach of his arms, whether a tree, a horse, a blackbird, or a river, had an existence equal to his own. Christian fairly gave off heat from his raiment when the sun shone through him. He was like one of Covington's brothers dyed in red and always running away; and grand in the mood of his Pa and brothers when the boy liked them best, as they grinned and tossed him in the air, and caught him roughly. He wanted Covington to follow him to the Celestial City that shone from a cloud farther on. It could have been London, that city, for most had never been to London to know any difference; London, where the buildings were sculpted in gold and shone with celestial ice.

Covington as a small boy felt happy inside the chapel where Christian strode in stained glass. Everything was newly made there, planed and nailed by sincere English carpenters and plastered by English artisans. It was done in the spirit of realness yet formed an other-world for Covingtons to take inside themselves, just as surely as if they had swallowed the mysteries of the Hindoo. A smell of freshness filled the boy's nostrils. He sniffed the grass under Christian's heels and heard the gurgling of a stream lined with buttercups, which Christian would leap next after he cleared the stile. That jump was said to be near Elstow, where Covington had often sat sucking a grass-stem and looking at clouds. He was able to hear the Elstow bullock low down in the next field as it challenged the smiling man; it made a grunting noise in the field—a sound that Covington made in the back of his throat, first being the bullock, then being himself butting the bullock in imagination and getting his first taste of joy from a fair fight.

'Come along and be quick about it,' Christian seemed to say, even before there were any words in Covington's head or ability in his legs to jump along to a command. Fixed in his blood from that earliest time was a readiness to respond to a beckoning gesture; and later, when that gesture was not offered, to boldly seek it and be sure it was made.

In the next part of memory Covington's mother let go of him and the man reached down and pulled him up into the wall. The congregation sang a hymn:

*He that is down needs fear no fall,*
*He that is low, no pride;*
*He that is humble ever shall*
*Have God to be his Guide.*

The sailor stood on the footbridge overlooking the canal and stared at Covington in the water. Whether he looked with interest or just gazed in that direction like a blackbird with quick, sharp, alert-headed movements was of little account to Covington, who didn't like it at all. The sailor scratched his ribs under his red waist-jacket. He wore flared canvas trousers, and on his feet were wooden clogs. His hairy shins showed bare. The sun glittered on the water and blinded him. He put a hand to his round beard and bunched it in his fist, giving it a twirl. Though it was the dress of a sailor he wore, the canal was far inland from the sea.

Who the man was Covington would learn when he saw him again in autumn, and remember him as if he had been planted in his brain and stored there to ripen. It would be cold by then and John Phipps would wear an overcoat and a cocked hat and call to a crowd under a lime tree with words to twist a rope around Covington's heart and haul him up from being down:

*I am content with what I have,*
*Little be it, or much:*
*And Lord, contentment still I crave,*
*Because thou savest such.*

That summer day, however, the man stared into the canal a long time; too long; and Covington made a blurting sound like a wet trumpet to accuse him of foul curiosity. When the man still stared, Covington grabbed himself between the legs and gave a tug and yelled, 'I caught a fish, it's a big 'un, look, see?' Then Covington saw the man tiredly grin and heft his sack of belongings across his shoulder and screwing his eyes against the glare of the sun disappear from sight.

The day emptied except for hens from the lock-keeper's cottage giving themselves dust baths on the towpath. Covington climbed from the water unpeeling strings of green weed from himself, giving a shake like a dog, then mopping his chest with his clean shirt taken from his bundle that smelled faintly of beeswax from Mrs Hewtson's understairs cupboard. He dressed and was cool, and was clean enough, too, but still carried the over-sweet fatty odour of the slaughteryard about with him. It would never go away for as long as he lived in his father's house.


Sunday meant chapel, where Covington sat next to Mrs Hewtson. She was his plump excitable stepmother, a fresh-faced widow and the best friend Covington ever had in the world. His real mother had died leaving him with a memory of sweetness and a green ribbon his father had placed in their Bible. Mrs Hewtson wore her best Sunday bonnet, which Covington told her was pretty, and she said he had better not say that to just any maid, or she would be jealous. She had rosy cheeks, humorous eyes, a teasing kindness and a great devotion in her heart. He played the fool and ground his knuckles into his forehead and dribbled spit from his hanging-open mouth onto the bare dusty boards between his boots. Mrs Hewtson nudged his knee and giggled, offering her bunched handkerchief to wipe his lips, and whispered, 'Be serious about you, now.' From a low, sinister angle Covington flashed his pale blue eyes at her, smiled and grunted, going at her with a small jerk of his head just like that bullock. She was very young.

'Stop it,' she squeaked, and the preacher, Mr William Squiggley, paused in his delivery, sending Covington a look of accusation: 'What was the last thing I said, Simon Covington?'

'You said that Abraham heard the voice of God and he took his son into the desert.'

'Why?'

'To cut his throat, that's why.'

'To make of him a *sacrifice*. And what happened next?'

'You did not say what happened next.'

'That is true,' said the honest Squiggley, who was their printer and bookbinder in his weekday trade, and easy with bad debtors because they were all good Christians and true.

Squiggley continued the story of Abraham and how there was a ram caught in a thicket, and how the life of the animal was taken for that of the boy about to be sacrificed in obedience to God. Covington lifted his eyes to the window-glass where he lived in his thoughts. His Pa dozed, dreaming of Mrs Hewtson's bezooms that were like jellies in his palms when he woke in the mornings. The brothers, matched each to their future wives in adjoining pews, had the look of dozing horses. Their ears twitched when the preacher's voice rose, and steadied when he prayed.

Covington raised his hand to answer a question about the boy, Isaac, and how he must feel being released from having his throat cut. He did after all go forth and father the people of Israel, and nobody else seemed to know that. But the preacher wanted no more of him.

Come weekdays Covington sat at a high desk in a leather-merchant's loft where he copied letters and entered transactions in ledger books. He had none of the scuttling resentment and affronted secrecy of the older clerks, but gave all to his work. He thrust his tongue between his teeth and twisted his body in keen concentration, half-slipping from his stool and balancing himself on his elbows as he wrote, one foot toeing him from the floor.

Cattle hides came to Quentin House from South America packed in bales and tied with hemp. They arrived in barges up the River Ouse and were unloaded at Great Barford. Broken open in the warehouse they were dry as parchment and so hard that being slid out by warehousemen they made the sound of a shovel being scraped on stone. They stank with an odour of dried blood and arrested decay. It was a stink known to Covington before he was ever taken on as a clerk, for it hung about the slaughteryard of his Pa. Thistle-heads squashed flat were often found in the packages, and once a greasy-handled Spanish knife that was passed about and then wired to the wall in the front office as an exhibit denoting the romance of cattle on the estates of La Plata.

It was a great illusion of power, sitting high above the busy town with the economy of leather radiating out from under Covington's fingertips. Written words, with their dangling tails and spiny longitudinal bones, engrossed him as they flowed from the tip of his nib. He could go all day on a folio of long f's and deep y's. Bootmakers, jacketmakers, upholsterers, saddlers—all, down to the man who made leather stops for musical instruments and up again to the one who packed cushions for the royal coach, placed their bids at the auction rooms and came begging Covington's master for terms to pay, which Covington conveyed back to them in his fine copperplate that he had learned in dame school from the age of seven.

But when he heard his Pa say that his boy would one day rise and stand equal to the Quentin House in money and fame, then Covington felt his stomach shrink. Those Quentins were mean procurers of shoddy advantage. They were Established Church and looked down on Baptists and Congregationalists as less than thistle-weeds. 'No mind, they have found you good work, and a lifetime's employment to keep you away from sticking a knife into gore,' insisted his Pa. But there were nights when Mrs Hewtson's sprogs pulled the bedclothes from Covington in their sleep, and he rolled to the floor lying groaning and looking up at the stars through a small breath-fogged window, believing he was always to stay down.

Covington would spit his shame out at his work by grunting like a bullock, extending his leg into the aisle of the office and tripping the messenger boys who came running past. While they were sprawled gathering their wits he took his aim and dipped a chewed-up ball of paper into a dish of ink and sent it flying from the tip of a wooden ruler. He could give a boy a wet black eye and send him howling in confusion and be poised over his next page of invoices before the ruckus began from on high, and a culprit was sought by the overseer. Covington beamed his innocence back in the face of any accusation. Though later there would be a challenge to a fight with bare knuckles along the canal-side, and Covington would find that the boy he chose to bully had great spirit, and wouldn't give up, and so Covington wouldn't give up either, and they would fight down to the end, slugging, mauling, damaging, until their skin rubbed raw and they powder-puffed to the finish.

One day late in the year Mr Timothy Quentin, brother of Covington's master and a man with the manner of an undertaker and with a foul breath besides, asked Covington and six other boys to come with him to his rooms and be given something worthy of their services. The boys jostled to be first in line and the one with his hand out most promptly was Covington. He was given two dull florins and told of an excess of hides on the market. There were just too many cattle on the plains of South America and other houses were stealing the trade. What this had to do with Covington being rewarded he was slow to perceive, and only understood when he walked out of Mr Quentin's rooms, and found warehousemen stacking the clerks' stools and desks away. It made no difference that Covington was the one highest-praised. The busy room of boys and penmanship was to be made a storehouse until prices rose, and then the Quentins would have their hides as cheap as anyone. At the prospect they could barely hide their glee.

C arrying a half-eaten apple and a beef bone that was to be his dinner, Covington walked through chilly damp cobbled streets where houses leaned over his head and almost touched. He sat under the bare-limbed lime tree in Bedford town square, blinking around him at the unaccustomed hour of noon and seeing how worry and care seemed to line every face. He wasn't hungry. A gangling youth walked around calling salted pilchards and an old woman dragged a bucket of slops between her knees and when Covington tried to help she scolded him. What was he to tell his Pa? That his sire's pride was just a nag to the slaughterhouse, unwanted, without value, scorned? That same morning Covington had whistled and thrown conkers at stray dogs and everyone had seemed to be laughing in the brisk smoky chill. Now faces looked pinched. The same youth returned and called his salt-fish with his head thrown back, shaking his tray, and they were slow to be gone at a ha'penny a clutch. Covington dropped his head between his knees and closed his eyes. He did not know how much time went past. He was in the Vale of Despond.

But there came a sad moaning in the air like a swarm of bees beginning its flight. He listened without raising his head or opening his eyes—only his mind came to attention. Then curiosity overcame him, and he blinked and tipped his cap behind his ears and looked around. It was not bees but the sound of a song coming from a huddle of men in the square. One was the salt-fish boy. On looking closer he saw they were *all* boys and not much older than Covington himself. They had weatherbeaten long faces and a look of the earth about them, as if they had climbed from sleeping in the ground. He had seen them beforehand, separated from each other, smoking their little clay pipes and scuffing their poor boot-heels in the shiny, overtrodden ground. They had seemed like anyone else he might see that day, resigned to a change in their lives that would never come. Now they formed a line. They had peculiar life in them. Covington's spirits gave a lift. A man joined them, older than the rest by far, and Covington recognised him with a kind of longing excitement in his heart: he was thin-faced, curly black bearded, and wearing a cocked hat and a sea-captain's overcoat that swished the tops of his boots when he made his determined stride. He had famished red lips and an excitable smile. One minute he had not been there; the next he sprang from the pavement just a few feet from Covington's face. 'That is my sailor,' thought Covington, 'who always goes round staring at a body.' The sailor took out a jew's harp and sounded a key of C. His companions broke into a shanty:

> Brace up the yards and put about
> Cut a fine feather and fly
> Give her a foot, she'll go like a witch,
> Sail till the seas run dry.

Covington jumped to his feet with a look of bright amazement. 'I'll be in this,' he muttered, and ran with others to where the quintet performed outside the baker's shop, their arms around each other's shoulders and their boots kicking right and left. Covington clapped his hands and shouted 'Oi!' at the end of every verse:

*The King's commission is all we need*
*To climb the rollers high*
*Eternity's port on the other side,*
*Sail till the seas run dry.*

'Oi!'

*Sail till the seas run dry.*

The baker came out and handed around sweet buns. Covington took one and sank his teeth into it. Then they went around the town venting their chorus on whoever cared to listen, stopping on corners, being handed more food, gathering coins. Covington went with them for a good few hours with all the fascination of a stray dog yelping at the moon. One of his brothers found him, and said they all knew what had happened at the Quentin House. They were sorry. What would he do now? He had not considered that, except that he was doing it, and his brother clapped him on the back, said it would do to warm the day, and let him alone.

In his uncracked soprano, Covington sang as joyfully as anyone, learning the words as he went along, adopting the rolling gait of a sailor and obeying the signals of the leader, John Phipps, the sharp-eyed seaman who at each turn when the boys formed a square raised his arm and fluttered his hand like a flag in the wind, laughing, smiling and encouraging the dance.

More than once Phipps caught Covington's eye. More than once Covington laughed back at him. John Phipps was a gamecock challenging and strutting.

Then Phipps stopped still and said to Covington directly:

'Do I know you, boy? I think I do. I think I know your heart what's more.'

Covington dropped his eyes from being known. He felt a nakedness to be covered, and nothing to shield him.

Then the seaman slipped his jew's harp to his mouth and cupped his hands around it, making a tune that slowed everybody down, and brought them breathing slow and feeling warm and happy into a circle around him. They were back in the town square again. There came a last twang that faded into the silence.

'Be still,' was the meaning of that signal. 'Furl your topsails and drop your anchor.' The sailor called for his squadron to kneel and be given a blessing. The crowd that had gathered wrapped its rags around itself and shuffled in a little tighter. They were the poorest of the parish, hungry for dreams, and if they could not have their dreams then toss them a sugar-crusted bun, and if the baker was not inclined to redouble his whim, then give them a pilchard from the fishboy's basket. Give them something. Even words to chew upon.

John Phipps gave them his sermon. He said he knew a great admiral, and the admiral was Lord of the Fleet. The admiral was one in a thousand and could do many things at once: he could build ships, launch them, serve in them, sharing travail with their sailors, and he could fight with them when dying, leaping from ship to ship and always being there with them. Yes, John Phipps had met him and knew him. Yes, he had fought alongside him as his Lord's Obedient Servant and had seen him die. And behold in the morning of the third day after the battle-smoke cleared, had seen him with his very own eyes, a man brocaded in gold and wearing a hat like a crown and carrying a book in his hand that was the King's regulations of truth writ on his lips.

'How can he live when you saw him die?' asked a beggar, with hope.

John Phipps bent down to him.

'Landlubber,' he said, 'for the love that he has in his King's service, he is sure in the world that comes next to have glory for his reward.'

He took a testament from his pocket in the last grey dusk. Resting a foot on the worn roots of the lime tree he struck an easy pose, throwing his coat-tails back. His intensity had a hunger to it, Covington saw. It demanded everything to itself. And when his audience listened, as he bade them to, and only sparrow-chirp and the grind of a passing cartwheel disturbed the silence, the hunger disappeared. Phipps's feverish eyes and his pained smile gave over to a changed appearance.

He caught Covington's steady eye watching, and asked, 'Ain't that right, boy? Ain't that how we can die and live?'

'But isn't *he* the King,' Covington asked, 'if he can do all them things, jump across water, live again, come back on the third day, and all? Isn't he the one who rules everything,' and added stoutly, 'isn't *he* the King of the Jews?'

'A clever boy who exceeds my parable,' said the sailor, putting his arm around Covington's shoulder, 'says my Admiral is Jesus of Nazareth and indeed he is. What an emissary he makes. Yet though he is called King and Master,' (here twisting Covington's ear with sharp humour) '*I* call him Admiral. He is the only man whom the Great King on High has authorised to lead the fleet in which any of you may serve. Wherefore take my meaning. Bear in mind my parable lest in your journey you meet with some that pretend to lead you right, but their way goes down to death.'

It was almost dark and Covington's supper would be on the table. Mrs Hewtson would be sure to whack him on the head with her wooden spoon when he came in. But he lingered and heard the boys say it would soon be time for them to get back to the place where they would spend the night. Though the boys were ready to go, John Phipps flourished his hat and placed it at his feet. A feeling hung about him like smoke. His black curls jiggled on his head like springs being constantly plucked by an invisible hand. He chuckled throatily, excitedly, with a promise in his voice. When speaking of his enemies—the mention of whose names caused his voice to tighten and rise in intensity, and the tip of his pointed Adam's apple to tremble—he made Covington feel that whoever John Phipps hated, then they were the ones Covington hated, too. Among them were ships' pursers, weevils, bishops, landlords, hoity clerks and all enemies of the poor and needy, and those who refused pilgrims their barns to sleep in. Also those who drove pilgrims from their natural estates, denying them the animals of the earth to grill in their fires.

'You must hate all of England, then,' said Covington, 'if that is the case.'

The sailor turned to Covington again, switching from ferocity to that look of quick good humour that Covington saw in him at their first coming across each other:

'Are my enemies those who lie in sluggish water, and think their sluggish thoughts, and make mockery of heavenly desire by carnal mimicry?'

Covington dropped his chin, feeling whole as a child in the company of this man. 'They are my *sins*, I confess it.'

'Nay,' said the sailor, ruffling Covington's hair. 'Some would call them sins, but I would not—since you own them so freely.' He turned to his boys. 'What do you say, lads, shall we have him in our fleet?'

John Phipps's four adherents all said they would, but it was not their selection that counted, they added, it was the Admiral's word.

'Well spoken. But I think the Admiral would have him in his fleet any day,' said Phipps. 'He has goodwill for boys.'

'Then heave-ho,' said the others, grabbing Covington by the shoulders and frog-marching him ahead of them down the road. With their meeting over they declared their starvation, and broke into a trot. Covington ran with the sailor and his boys through the dark, out on a muddy road past the town and into a barn where they made a fire of sticks. They strung lanterns on beams and roasted potatoes and turnips in the coals. Covington liked the way they did everything with a snap, a rush, and then stretched their legs out before them and smoked their fierce pipes, which they plugged with tobacco handed around in pinches by John Phipps. One of them produced a pullet from his cloak and made ready to despatch it to perdition, only to find John Phipps's cane across his neck. He asked where he had got the bird, and only allowed him to strangle it when satisfied it was from under a bush near a yeoman's farm. 'I would have *you* plucked if it came from any deserving poor,' he said.

With firelight licking their pinched faces the boys told stories of where they were from. All but one belonged in Bedfordshire and neighbouring counties. Like Covington they had been ejected from their workplaces or else had never known anything except wandering the roads. In the right season if they were lucky they dug potatoes, cut willows, and drove turkeys to market for the reward of a few grubby pence. Now they would take their chance on the sea. None of them except Able Seaman John Phipps had ever worked the sea, but it was their sworn intention to do so. Indeed, as Covington soon learned, such was the whole purpose of John Phipps's preaching—to take boys with him to the ships. It was to cultivate and escort to the naval yards of Britain a clutch of would-be sailors imbued with a parable of Christ which they would live-out in rough waters. For what was a Christian to do except bear witness to his fellow-man, and if driven to extremes bear it alone where there were no spectators, on the perilous deeps. There was no better test than that of a Christian's mettle. But Phipps was not in a mood to lead his boys to Portsmouth in a hurry and find them Christian commanders, of which he knew several. He first wanted to check they could read their scriptures, and show in their hearts a love of the unseen. Then they would be a power on the four seas, and return home with treasure beyond reckoning.

Covington did not know if he had a love of the unseen.

'It was given to you by nature,' said Phipps.

'What is it, though?'

'It knows you. Pray stick by my side. You and Joey Middleton here, I think you are my prizes.'

Joey had a small, sad and eager face. He had the sniffles and a runny nose, and wiped his upper lip with the back of his hand. John Phipps gave him a woollen comforter from his deep pockets to wrap round his neck and keep himself warm. Joey was eleven years old but looked younger. He was the only one who knew ships, being a West Country boy from Devonport, where his father, he said, was a sailor with red hair. And his mother? She was in that town, too. But that is all he seemed to know, and John Phipps said that he had found him on a scow near Hull, curled up on the deck as if he were chained there. A lean bosun and his wife had taken him in charge. They claimed he was their own, whipped him when they liked, and used him as their lackey or galley slave as they made their way around the coast. They had got so far from Joey's birthplace that he believed himself to be in another country altogether, where English was barely spoken. And he forgot that he was free, had no conception of prayer, and so was in a fair pickle when John Phipps stole him away and started breathing faith into his bones.

By the flame of a candle the Book was passed around. All stumbled over the words until it came Covington's and Joey's turns. The two outshone the rest in a reading of the Prodigal Son. It had a special meaning for Phipps, and they guessed he was estranged in the way of a prodigal himself. It was then everyone's bedtime and they heaped up the hay. John Phipps would not let Covington stay, but sent him home in the moonlight with a promise that he would call upon his father in the morning at the slaughteryard, and parley about Covington joining his boys.

True to his word John Phipps came there, and put his case.

'If you mean my boy to be an evangelist, like yourself,' boomed Covington's Pa, rising to his full height wearing a leather apron, addressing Phipps and waving a willow-stick around (that he was using to beat horsehide, to loosen the hairs from it to make head-plasters), 'then you have chosen the wrong boy. My Syms could no more persuade a sinner from off his path than that sparrow there,' he pointed with his wand, shooing a scrawny bird a few hops away, where it splashed through a cesspool composed of blood, chaff and urine. 'He would as soon dirty himself in sin as cleanse himself in the rivers of Babylon—'

'If you mean he doesn't know what sin is—'

'Aye, I do mean that. He would willingly serve the devil for a pat on the head, and likewise raise Christ's hem from the dirt, e'en if it skunned his knees to the bone.'

'I can see that in him,' said John Phipps. 'It is why I want him in our crew.'

'Yours is a daft ship, being on land without keel or rigging,' observed old Covington, with a barb of suspicion in his voice. 'Are ye supported by any missionary society?'

'Only what God provides.'

'And your nose, from the sharp look of you.'

John Phipps smiled. 'You shall have one less mouth to feed if he comes.'

'That has occurred to me,' said the father drily, 'since yesterday, when the Quentins lost their market in hides.'

'I want to go,' said Covington.

'It is to my sad advantage to let you go,' said Covington senior, 'whether I like it or not.' He grabbed Covington and pulled him to him. 'God love you, lad, as I do, and you'll meet no harm.' Then Covington went to Mrs Hewtson, and she grabbed him to her too, and said the same kind things to him. 'What shall we do without you? Who will bring us such cheer?' Word meantime was sent to their preacher, the printer and bookbinder Mr Squiggley, to ask for information in the matter. He said that John Phipps was known thereabouts; he was a wanderer; he had formerly been a rogue; his father had disowned him; he had seen a better way; he was known for his eloquence with boys. So really there was not a bad word said about him except by those who feared his scorn. His temper was fierce, he was possessive when roused, and was the special hate of gamekeepers, on whose land he trespassed in a lofty spirit of freedom. It was said that he set a pace with his fast legs like Alexander the Great in crossing England, and as a boy had sailed in the English fleet against Nick Frog, and so was branded young to the ways of the sea.

Mrs Hewtson stuffed a canvas satchel with rice pudding, cold mutton, cheese and bread. She handed Covington a bottle of mulberry wine which he slipped in his pocket. There was no more room left in his satchel. It was heavy, almost unbalancing him as he slung it around his back.

'I am a packhorse,' he said.

'You are a donkey,' jibed Mrs Hewtson, tugging his ears, putting her arms around him close. 'John Phipps's donkey and he's got you cheap, my darling heart.' Covington gave her a hot, tearful kiss. She had been long enough in their house for Covington to have forgotten when she came. All the busy, cosy, forgiving and playful times of his life were spent within a whistle of her arms and lit by the shine of her oven door. Nights of being squeezed around the hearth, scrapping and boasting in a parlour the size of a thimble.

After leaving Mrs Hewtson he went round to his brothers, one cuffing his ears, another pulling his hair, a third booting him in the backside or, as he worded it, 'giving him instruction on how every fat must sit on his own bottom.'

It was his launching into the world, where he believed there was nothing to hurt him unless he procured it for himself—though he was a little pained by the ease and convenience with which his Pa and Mrs Hewtson now let him go. Just as he had been marked for a trader in his Pa's eyes, now in this turnaround he was marked for the sea.

'When you get your ship,' boomed his Pa, 'then I believe there will be no stopping you.' He turned to Mrs Hewtson, pinching her cheeks to cheer her up. 'Why, my little butter-churn, our boy shall one day have his own ship to command—our boy shall be Nelson, Drake and Dampier, all three in one.' When Covington's Pa boomed praise he was heard for a mile around, and 'twas often said of him as a horse butcher, that he was a great hoarse as well.

# BOOK

*On an Art of Bumpology*

# 1858

It was the hottest time of the year, a month after Christmas at latitude thirty-five degrees south. All down the New South Wales coast columns of smoke rose from fires burning inland. The unchecked flames shot their smoke in the air forming anvil-heads of cinders. At night the fires burned low to the ground in a lurking, underhand fashion, bothered by sea-mists. Next day they flared tree-high again, greedy, fed by hot winds from the parched inland. The fires had the sniff of rage about them. The sea was the only barrier to their eating a man's face off. Sparks flew out over surf as tongues of flame advanced onto headlands. Ash fell in the water and darkened the white beaches.

At Tathra, in the far south of the colony, Mr Syms Covington embarked for the port of Sydney, as was his custom every six-month, on the schooner *Skate* from Twofold Bay. The voyage of two days was done in a haze of burning. Covington had a good stomach for the sea but was unable to sleep. He experienced cold sweats and a discomfort that pierced a sword to his belly. He stood on the deck of the *Skate* watching worms of fire in the hinterland and knowing there was something wrong with him that a swig of gripe water and a good hard belch would never fix. He crouched in a chair, pulling his knees tight against him, and then stood clinging to the rails. He lay down on the

deck and was no better. The captain prodded him with the
toe of his shoe. His condition made him afraid. They sailed
north, pitching and rolling. His battered, broken-nosed face
turned square to the wind had the look of an old prize-
fighter's coming up to a bout.

Entering the Heads of Port Jackson just after dawn, the
captain found Covington utterly stricken. His eyes were
open, watchful, but he uttered not a word. With sails slack
and the schooner steady on the tide the sufferer was
offloaded forthwith and rowed to a Dr MacCracken's
cottage in an arm of the harbour at Watson's Bay.


As MacCracken first saw Covington he was the colour of a
ripe plum, barrel-chested, massive in thigh and limb, and
silent as the grave in his agony. Covington was then forty-
two years of age. His impressive head rested on a folded
coat. One fist was clenched, and when MacCracken prised
it open he found a small cone-shaped shell with four valves
at the top. It was a common barnacle and he threw it away.

'Get him to the house. And hurry.'

Men carried Covington on planks to MacCracken's
library and he prepared his knives. 'Get sheets,' he yelled,
'and spread them around the floor.' The last consideration
in MacCracken's thoughts was the saving of a life, for he
believed the man as good as dead with advanced peritonitis, but not last in his actions, you can be sure, which were
swift and useful.

Covington parted his eyes a slit. Nothing else in him
moved except his eyeballs, which followed MacCracken
around the room. He observed that his saviour was a young
man, lean-necked, tall, vital as a whip. He held his lancets
and scalpels to the light, and drew them across his thumb-
nail to test their sharpness.

MacCracken kept himself calm. He had no great love of
surgery, indeed had only recently begun in that business

and doubted his wisdom already. Yet his hands were steady and more to the point he knew that if such advertisements for his skill as this Covington had ample pockets, then so much the better. For MacCracken fancied soon to select himself a slice of that wide-open land of Australia where he could put a man to manage livestock, and so guarantee himself regular percentages without having to dirty his feet in dust. It was how fortunes were made here if you were wise enough, and better than gold. And so was Covington his godsend? You may be sure he was.

A pall of bushfire smoke rolled along the coast and suffused the harbour foreshores, entering the room where the patient lay and stinging the surgeon's eyes. Without delay MacCracken administered chloroform and put Covington to the knife, delivering him of a foul, instantly bursting appendix. A mere half minute divided the patient from life and death.

Covington blinked awake and found himself among the living. But which lot of people and where?

'Don,' he croaked, and reached out a crippled hand.

Where that 'Don' came from MacCracken had no idea, though it declared a bond of vehement familiarity between them that was to last.

Say there was nothing between them at first except mistaken identity (who was this 'Don' at all?), and then that a quality thickened in the air between them—like a lens they could use to know each other better—one man adamant in his being, that man being Covington; and the other, the younger, MacCracken, with his limp brown hair and bony nose, ready for wisdom without having a clue that he was.

The first time Covington spoke, MacCracken learned he was deaf as a mountain. His cheeks needed a good hard pinching. 'Wake up, old dodger!' But yelling did no good unless made hard against his ears.

'I had a shell!' Covington shouted in the half-light.

'I threw it away,' said MacCracken.

'Where is my shell?'

'*Gone! Vanished!*'

'Mind the reef!' Covington shouted.

'Mr Covington,' MacCracken held him by the shoulders, '*you are on dry land.*'

'I had a shell!' (etcetera).

MacCracken flung wide the curtains. It was barely surprising that in his delirium Covington believed himself aboard a vessel, considering the fine chronometer Mac-Cracken had on his wall and the proximity of sea-water breezes wafting through the window. There were books on tables and spilling from shelves, many with a nautical flavour, and in a corner alcove was a fine globe of the world of the sort favoured by ships' captains. MacCracken rented the house from the widow of one.

Covington narrowed his eyes and looked at his saviour with a cunning suspicion. MacCracken looked back at him lazily, now. He was an American on his way around the world from Boston, having come to rest in Australia after trying the gold rushes and exhausting his sense of adventure.

Covington began to struggle again. 'Don?' he barked in his delirium. MacCracken wrestled him down.

'The name is MacCracken. You are under my care.'

'Don *Sia Di*?' Covington said, or so the name sounded to MacCracken's ears.

'David D. MacCracken is the name. Just as I said.'

Covington wearied with repetition of his 'Don', coming out of the influence of chloroform, that thin colourless liquid with an ethereal odour and a sweetish taste with which the new-made surgeon had stilled Covington's strug-gles—and sometimes, for the interest of it, had enhanced his own senses and coloured his dreams by taking a sniff.

Finally MacCracken shouted against Covington's ear and his meaning won through. 'I am your doctor! You are ill! Be satisfied!'—and Covington sank back in his pillows, making a dry chomping sound and rubbing his battered nose with the back of his hand, giving MacCracken the benefit of a gentle smile, which the younger man witnessed then for the first time, and it warmed his liking.

'You're an American,' said Covington.

'You thought I was someone else?' mimed MacCracken.

'Aye.'

'A Spaniard?' snorted MacCracken, snapping his fingers, clicking his heels, doing a fair tarantella in charade.

In time MacCracken would learn that the man Covington called him in his delirium was also nondescriptly brown-haired, also big-nosed, also obliging of manner, also absolutely unremarkable-seeming and doubting his first-chosen trade, and aged but thirty years the last time Covington saw him. No Spaniard, either, but a well-born Englishman, and around six feet tall and so inclined to stoop a little in his relation to others. His name was Charles Darwin but MacCracken was far from knowing that, and would have thought it unlikely even if told, Darwin being famous for his *Beagle's Voyage*, which MacCracken had read at the age of twelve, holding it somewhat responsible for nurturing a whim, that bore fruition, for science and travel.

'I am sorry to give you this trouble,' Covington said, coming round in a cold sweat.

'Not at all,' responded his saviour.

Under wiry eyebrows and a clifflike forehead Covington's eyes followed MacCracken everywhere as he cleaned his instruments. Covington was a powerful presence in the dim light, the planes of his cheekbones and jaw offering a fine portrait. MacCracken was interested in his head. Lumpy, he would say. But interesting.

'Your hatmaker,' he supposed, 'finds his fortune in you, Mr Covington?'

That head's resemblance to a loaf of bread, where yeast pushed the crust in various stern directions, had often been

remarked upon with Covington. His ears hung a little pen-
dulously in his age. MacCracken, with a flippancy to his
nature, muttered whatever he liked while in Covington's
company, never expecting a reply unless he bothered with
shouting. Covington's hair was thinning and black and, 'I
daresay dyed, old fellow?' said MacCracken, testing the
emptiness of the air.

'Blustery weather,' Covington replied.

Covington's facial purpling came from old scars.
MacCracken used his magnifying glass. He deduced they
were powder burns but Covington said nothing. Facial
scarring was not the only mark on him. There were welts
on his shoulders, embedded like sea-slugs, purple and slack.
He guessed that Covington had once been severely flogged,
and from turns of phrase Covington used ('deaf as a mainmast' and 'sparm fish' for whale), divined in Covington's
distant past a ship, though whether a merchant ship, a
convict ship, or a man-of-war he could not tell.

'What ship? What navy? What crimes? What cruelties?'

Covington gave no answer.

It was the nineteenth of February by MacCracken's diary, and Covington had been with him twenty days. 'I am weak,' Covington rolled his eyes around. 'Will you care for me, MacCracken?'

'I am doing so already, crippled old dog,' the young doctor murmured, conveying kindness by giving Covington's arm a squeeze. It was not MacCracken's intention to run a hospital for his cases, but with Covington he heard himself prattling: 'Of course, yes, rely on me, sir, I shall make arrangements, etcetera,'—all condensed into one shouted word in his charge's left ear (the better one): '*Yes!*'

With an instrument sent from Boston by an old professor who still had hopes for him, MacCracken tackled Covington's ears. Gobbets of wax blocked his view. After careful syringing he saw that both drums were scarred beyond recovery. It was as if firecrackers had popped inside them. Covington's submission to his care was touching.

'I went to an aurist about this,' Covington tapped the side of his head, 'and he said for a thousand pound he would cut me open and clip my ear-bones, and maybe I would hear better. Would I?'

'Keep your thousand, grandfather.'

Mr Covington dozed. MacCracken felt a protectiveness towards the old coot as for a gruff, well-meaning peasant with a crock of gold. A man who could spare a thousand like that would know of some prime investments. Trying another sort of examination MacCracken ran his fingers across Covington's scalp. It was like playing on a bag of stones, and using instinct aided by phrenology (at which MacCracken prided himself, believing the craft to lie somewhere in the direction of a firm prediction), he sneaked a mental picture of Covington to verify his first impressions.

The message MacCracken read through his fingers came to him in a few moments: a doglike fondness was no surprise; the potency of an old sire; powers of concentration and challenge; a streak of resentment; the capacity to deal damage; a certain helpfulness; secretiveness.

This last was no surprise.

Covington came awake as MacCracken felt what he had once heard called the 'band of hopefulness'. It was ridged across Covington's dome, a veritable rainbow of potential joy, and not seeming to belong with the doleful stranger at all.

'What are you doing? Are you "reading" me, MacCracken? I won't have it!'—and he thrust his examiner's arm aside. 'You won't *use* me?'

'Dear Mr Covington!'

'Bumpology. I *spit* on that art!'

'Mr Covington!' (louder in his ear).

'Yoi?'

'I—am—your—physician.'

'You—are—my—meddler.'

Though Covington gave a quick smile to cover his outburst, and MacCracken smiled with him, they both were astonished by the vehemence of the exchange.

'Pardon me,' Covington said. 'I had a bad time with that business once. When I was *jugged* and *bottled*.'

'You are pardoned, sir. When was that?'

With the shimmering half-understanding the deaf have, that is also like a charm, Mr Covington scuttled back inside himself and secured MacCracken's fascination with that 'bad time' and that 'business' by keeping his jaw firmly clamped. It must have had a good outcome, surely, thought MacCracken, because the rainbow ridge of hope said so. Either that or Covington's fate had not yet run its course.

Covington lay on a bed in MacCracken's library and gazed at MacCracken's books, read their spines and threw his host a sprat of information to chew. 'I've come home, it seems,' he said. '*Home*,' giving the word a scornful edge. He named a few titles—Murray's *English Grammar*, Mackintosh's *History of England*, Byron's *HMS Blonde*, and Darwin's *Voyage Round the World of HMS Beagle*— saying he 'owned those too', which MacCracken thought, at the time, a pretty ripe boast for such an old carthorse.

'It is like a ship's cabin in here. I like it very much.'

Through several days MacCracken watched Covington closely for signs of relapse, and one evening, having moved him to a bare side room for his convalescence, witnessed another responsive quirk in the man. Covington reached from his bunk and touched the walls with his hands. Splaying his fingers he pressed his palms flat in all their sweatiness. MacCracken thought he merely craved the coolness of stone, but learned (in time) it was otherwise with Covington. For deaf as he was, Covington held his body taut as a tuning fork, and listened, and heard—for the world sang to him through the sounding box of 'Villa Rosa'.

Much later MacCracken was to learn all that the walls meant:

Touching them brought back to Covington his adventures, beginning with his earliest on leaving home at the age of twelve. It was the suck and slam of the ocean, the great stringed instrument of wind Covington detected, combing through eucalyptus branches overhanging the slate roof and sifting him down his only music in small vibrations, the hard thrum of cicadas and the decisive slap of small waves on hard-packed sand. The creak of a ship's timbers, the rush of waters along the leeward railing and mashing across half the deck like a neverending turnpike. Then the dip of paddles in a quiet estuary (on his several returns) that said, 'England at last!' The wetlands, the flatlands, the stink of mud and rutted roads. The hiss of footsteps across dewy-damp grass. Empty houses. Graveyards of names. A door-hinge creaking as he entered an old chapel, and then with his eyes lifted discovering that a window that was formerly there, high in a wall, was gone, and some greatness in his heart leaping the obstacles of the world was gone as well.

When Covington saw MacCracken watching, he sneaked his hand under the covers, embarrassed at showing his feeling.

MacCracken had this heavy-limbed Ulysses in his household care for another week, and then—for Covington remained feverish with a persistent infection where the cut had gone septic—arranged a cottage, 'Coral Sands', where he could attend him daily. Covington stayed there through all March and half April, well able to pay a good price. He retained Nurse Parkington, MacCracken's sometime assistant and a woman of ample spirit and powerful arms, to dress his wounds and, when he was much improved, to pummel his stiff joints while he sprawled walrus-like on a table.

One day Covington asked if he might call on MacCracken, convivially, he said, and, without much ceremony, the doctor found him at his door. Covington's hair was combed straight forward like Napoleon's, with a curl over one eye, and he reeked of pomade. His prickly devotion came at MacCracken from under a cliff of forehead, and he beamed his great smile, bellowing 'MacCracken!' so that his listener might know from his admiration that something was wondrous about him. In retrospect it did MacCracken good to feel the heat coming from the slab of Covington standing in his doorway. But in the present it itched him around the collar. It might be called love, that tide or spark of feeling the other gave off. When in later years MacCracken got through to a settled plan of life—and returned to Massachusetts to square his accounts with running away, and became, in time, a puzzled student and then a practitioner in matters of the mind—it was often this picture of Covington holding the door-jambs that recurred to him. It was an emblem he took into himself, indeed, as a measure of character. Never give up, it said. Neither your victories nor your losses. Stay eager for your pain until it serves you well. Nourish life to the end.

'You have made me good,' Covington boomed, producing a bottle from under his arm and holding it up in the air: 'Rum tiddley-um-tum!'

MacCracken had asked for this, and over a glass of spirits Covington confirmed himself as a case to be admired. He began quizzing MacCracken on money matters. MacCracken had saved his life, and Covington's best answer to that was to begin money-making for him. When MacCracken mentioned acreages and sheep as being worth more than 'accursed gold', Covington nodded sagely, owning that he 'knew a man' with four thousand acres and a fine house in a district with good soil and fair rainfall, and who might be amenable to taking a new owner aboard. Thus MacCracken saw his way clear to much leisurely scribbling and a changed background to life. In his present, callow state of mind the ambitions MacCracken entertained were literary—he was a would-be essayist. For that he wanted a good thousand a year. Too much—but when was enough ever enough for a pleasure-seeker? As he farewelled Covington that night he put an arm around him and gave him a warm embrace. They stood eye to eye in their tallness. Their beaming faces said 'good fellow' to the world while their minds raced, calculating their needs and adjusting their tactics to each other.

'I like you, MacCracken, but do you think that matters?'

'Indeed I do,' the younger man replied, confused by Covington's question and unsure if he liked him quite as much in return.

'What's that you say?'

MacCracken cupped his hands to the side of Covington's head and repeated himself at the top of his voice. '*Indeed-ah-do.*'

'"Indeed yah do",' Covington chuckled, mimicking the other's expression, and stomping off into the night. 'Well said.'

MacCracken stood in the dark feeling bothered. Talk of getting rich was all very well, but there was a humour in Covington that niggled him; a way of acting possessively— which is to say without respect—as if this little patch of heaven where MacCracken lived was quite Covington's own, under a prior claim. There was a studiedness about the man, giving MacCracken the feeling that while Covington was honest in his gratitude, he most of all wanted to emphasise that he was no man's lackey. If anything, that MacCracken was the inferior in their two roles. The sum of MacCracken's feeling, too, was that Covington had set him up like a row of skittles. MacCracken could not get rid of a feeling that Covington had not fallen into his sheltering cove by accident. Something about his being there was contrived. Yet how could a man do that, almost dying on your doorstep in the attempt? It was not possible to manufacture peritonitis, and time a grave illness to within a minute or two of death. It made no sense in any understanding of the world at all.

MacCracken stifled his irritation soon enough, however. He was niggled but not so high principled as to take the matter further. Covington's getting about in good health was a case of satisfaction to him as he needed his advertisements hale. For questions were asked in Sydney about whether young MacCracken was qualified at all, to which MacCracken indignantly flourished certificates issued under Massachusetts law, inviting inspection: 'At your leisure, *if* you please,' and threatening lawsuits on the matter. He travelled each Tuesday to the government hospital to assist Mr Vincent Crews, a man little short of a drunken blunderer, but with vice-regal connections, whose rare successes, MacCracken indignantly countered, were MacCracken's own.

A short time later Covington made a fine recovery, then went about his business in Sydney Town spouting MacCracken's good name, and collecting views of the doctor in return—thence embarking for Twofold Bay on the next sailing of the *Skate*.

'Coral Sands' and 'Villa Rosa' lay among the cottages of watermen, sea captains, pilots and fishermen. Watson's Bay was prettily situated on the peninsula called South Head, with sandy beaches and calm sheltered waters on one side of shelving, dramatically broken land. The Pacific Ocean, invisible but close, shuddered against sandstone cliffs a narrow quarter-mile behind the settlement's back. The staging was theatrical in the written opinion of MacCracken, who got down to his essays in his leisure hours. On the ocean side it was a good setting for *Lear*; on the sheltered side you could have your *Midsummer Night's Dream* if you wished, amid cabbage-tree palms and stately, red-limbed angophoras, that were called apple-gums in New South Wales though they bore no fruit—a typical confusion in the colony, whose botany and everything else was upside-down like the seasons—until the timber was cleft, and then the resemblance to apple showed.

Just a few short years ago these shimmering protected coves were one of the last great unmeddled-with portions of earth. Then with lightning-swiftness (compared with the time that had gone before), pink-cheeked high-principled naval and army officers made it their England's preserve. With utmost reasonableness they spoke their laws, setting up a gaol, a gallows, and a series of fine government buildings made of sandstone blocks. Such was Sydney Town. Windmills turned in the humid hot haze, grinding a damp sticky flour. Charles Darwin saw them when he sailed through the Heads of Port Jackson twenty-five years previously—their listless, hopeless sails. Dealings with the black people who were there before the garrisons arrived were first conducted in a tone of amused equality, noted the essayist, with gifts exchanged and comparisons made of clothing, adornments and human anatomy. The friendly contrasts were well understood as reflecting that both sides were matched in being human, and this was soon ironically demonstrated by the women producing little bastards, by which time it seemed there was a belief on the part of the natives—with little language, and much gesturing on their part—that they had met with a superior civilisation and liked its products of rum, treacle and flour better than their own raw produce of kangaroo flesh and bush honey.

There was not much adventure in the journey bringing the young MacCracken to Australia, although he sometimes boasted there was—colouring accounts of brushes with *bandidos* and the like. He had believed the great age of adventure was already over. Boredom and a breach of promise dispute propelled him from his Boston home and he went to Buenos Ayres, acquiring a dissolute's Spanish. There he heard about Ballarat gold and made his run for Australia. Others took the journey with him seeking fortune—sick at heart rounding the Horn, then stitching up the coast of South America as far as Peru to find another vessel. Thence across the rollers of the Pacific in a stinking overloaded bark. First landfall, and a hasty one, at Chatham Island in the Galapagos chain, where twenty turtles were taken live for the ship's galley, and the knowledge of FitzRoy's *Beagle* and Charles Darwin having called there toasted in Chilean amontillado. Leaking and wallowing the whole way, they made, finally, landfall in New South Wales. Then came the march to the gold diggings of the 1850s through stupendous heat under a brassy sky. Following that, failure on the diggings. And retreat to this haven.

MacCracken walked his clifftops on calm days, wading through prickly banksia trees and flat-growing hakea bushes with spidery orange flowers. He came to a lookout point and never failed to amaze himself. The eye of the Pacific stared into unfathomable space. It was a great emptiness and a mystery after the closeness of domestic life near to hand. *Consider London's Hampstead*, he wrote in an essay (never having been there):

> *with the roiling South Atlantic instead of the 'Heath',*
> *and no end to be seen on the other side, only a glaring*
> *bowl of blue studded with albatross, white-caps, and*
> *sometimes a ruddering Leviathan.*

MacCracken felt a little afraid when he stood on those clifftops. Sight of the sea made him know there were things he was yet to dare, and so the sea challenged him, and he was often worse-off for his constitutional—more dissatisfied with himself than was good. Men like Covington he reckoned had faced the sea with inferior trepidation: they simply did not allow it to move them, and were lucky in that. It gave the world sailors at least.

So MacCracken with his snobberies got away from the cliffs and hiked back to the cove he loved so dearly. It was a world unto itself where afternoon sea-changes of weather, called southerly busters, raced overhead, leaving the cove placid while whipping the waters of the harbour farther in. MacCracken liked the feeling better than the mish-mash of weather stirring Sydney Town a few miles around the shore, where they had a summery succession of hot, cold and humid, with hot prevailing after the passing of the rest in a furnace-breath of westerly wind bringing the smoke of the ever-burning bushfires. Here was a more idyllic setting for MacCracken's moods. Let his friends find him 'on location', as he said in the fashion of the day, whensoever they wished. Let the world steam over him while he hunkered down a little removed from its chops and changes, a privileged spectator on a tiled verandah, a glass of ale in his hand, and his kangaroo dog, Carl, an amiably deficient hound, lapping milk at his feet.

All was cosy. All was right. MacCracken craved no cataclysmic dramas. He had his surgery, saw a few patients for cuts and scratches, and the rest of his time he gave to friendship, dalliance, and his literary pursuits. Nightly there came a tapping at his window-pane, and he enjoyed the close attentions of a visitor discreet as she was willing— 'The older Miss X,' he told his friends, 'if you want a name at all.'

MacCracken's essays were his personal pride and his most abject failure. Inspired by Emerson and Thoreau, they observed nature in a genial fashion and toyed with philosophy. They made humour out of his travails. '*The bird was so tame I killed it with my hat,*' he wrote of his Galapagos Islands interlude. '*In experimental mood, I persuaded a tortoise to haul a rock.*' He did not think the essays were bad, but though he sent them away in scrupulously tied packages, nought returned to him but silence. His soul went six months across the sea and came six months back unenthused-over. If anyone asked why he refrained from publishing in the *Sydney Morning Herald* in preference to Boston and Edinburgh journals, he said, 'You know as much of colonial taste as you do of my particular style.' But the fact was that the *Sydney Morning Herald* would not have him either.

He showed an essay to Covington, who kept it an hour, then handed it back with a grunt.

'Mr Darwin was there before you.'

'Well, so he was, old bookworm, and the whole world knows it through his journal of the *Beagle.* I saw the birds that he shot, finches, hardly bigger than mosquitoes, some of them were.' MacCracken shouted into Covington's ear: 'Bang! It was done.'

Covington twisted his mouth disdainfully.

'*Obtained*,' he said, 'if you want the right word. Not shot.'

'What is that you say, miraculous old pedant?'

He spoke the words to Covington's retreating back.

'Obtained,' Covington repeated without turning around.

'Sir?!' pleaded MacCracken.

Covington hauled back a leg and took aim at a coconut that MacCracken had placed on a stone as a decorative detail, and sent it flying through the doctor's garden of shells with a well-placed kick.

'Finches!' The word exploded into the night.

MacCracken sighed, then went back indoors. For a minute he heard the coconut bouncing through the rocks down to the shore as Covington pursued it to destruction.

MacCracken pulled his journal of the *Beagle* from the shelf and turned to a page headed 'Ornithology'. He saw that Covington was correct and that 'obtained' was the very word used.

Another six months passed and Covington returned from his home in the south. It was October and changeable weather to the point of madness, hot as the equator in the morning, misty and chill with low rushing cloud in the afternoon. MacCracken looked up one day and there Covington was on the horizon, unmoving, watchful. He sat on a hired pony. Much had transpired between them during their separation. They had corresponded on money matters and through the facility of Covington's Sydney agent MacCracken already owned cuts in a number of Covington's cargoes: whale oil, timber, hides. They were beginning to show him a handy profit. So there was no more welcome visitor on MacCracken's rocky patch than the man he saw.

'Mr Covington!' MacCracken waved his hat.

Overlarge on horseback, Covington was a stone pillar awaiting a lightning bolt. His movements were a story of shoulder bones, rib bones, ankle bones and skull all broken in his youth, mended, and thereafter very sore. Those 'boans' ground against each other and were stiff, lending Covington his monumental manner when he attempted turning his neck.

MacCracken strode up the white track. 'Old fellow, how good to see you!'

'That woman who helped me,' Covington looked down from his hired nag, 'what was her name?'

Covington's gaze had all the power of a lament. MacCracken helped him down. They descended the landscape of rocky headland and sheltered bay.

That day MacCracken wrote in his diary:

*I love the old fossil, he warms my liking, it is a good feeling to know we are friends.*

Two years later MacCracken went back to the date and circled the entry in red. It was because of a letter Covington carried in his satchel but was no more likely to take out and show around than he was to strip his clothes and walk naked. It was a letter from Charles Darwin and it read:

*Dear Covington, I have for some years been preparing a work for publication which I commenced twenty years ago, and for which I sometimes find extracts in your handwriting! The work will be my biggest; it treats on the origin of varieties of our domestic animals and plants, and on the origin of species in a state of nature. I have to discuss every branch of natural history, and the work is beyond my strength and tries me sorely.*

Two years. That would be the length of time Covington nurtured his pain before MacCracken understood the story he carried in his bones, and how it ate away at him. Two years before Covington showed MacCracken the letter. By then MacCracken would, in a kind of by-product of shame, know what his own role was to be in the tale of Covington's life, and how to go about correcting his ignorance and bringing his friend through to his end.

\*

Covington stayed a week subjecting himself to the pounding of Nurse. MacCracken had leisure to examine him for
his present state of health, which was excellent save for
his rheumatism and deafness. Like a farrier re-shoeing an
old horse he gave Covington's ears another syringing.
Covington declared himself well satisfied, though he heard
no better.

'What made you deaf?' MacCracken shouted a question
he had asked before.

As before he got a deflected reply.

'Why is a worm blind?' responded Covington, then
answered his own question: 'Because it lives in the dark,
that is why.'

MacCracken rolled his eyes. They were back to their first
way of dealing with each other, with Covington excluding
MacCracken somewhat, yet demanding his attention—and
always on the edge of a withheld confession.

'MacCracken, can I trust you?'

'As your physician and your friend? As your business
partner? *Aye* on all counts!'

'How wise are you?'

'Wise enough.'

'What did you say?'

'I said *wise*,' MacCracken nodded and shouted, '*enough*.'

'You tied a string to a tortoise and made it pull a rock.'

'Thank you,' croaked MacCracken into his chin. 'So you
hold my own writings against me. Thank you kindly. What
a good proof of readership. Alone of all the faithful you
kept the faith.'

'Since we met I've been wondering about something.'

'We never met, we collided, old barge.'

'Is a man only ever to be as he seems? That is my point,'
said Covington.

'And a good one, ancient Diogenes,' remarked

MacCracken. If *he* was only ever to be as he seemed, then he would be a flimsy sort of a fellow.

They smoked their pipes and watched the shallow tide. When it retreated it printed white sand with mottled hearts of Port Jackson fig leaves.

'Must a servant be always—just a servant?'

'Jesus was a servant,' said MacCracken righteously.

'Who?'

'*Jesus.*'

'Nay, he was a master,' said Covington.

'Come up to the house,' said MacCracken, clapping a hand on the shoulder of his brooding friend. 'I have a fine old brandy from the Cape.' Covington stared at him uncomprehendingly until MacCracken made the 'snorter' sign with his elbow, and then he leapt up and trotted at his side to 'Villa Rosa'.

'What do you think the brain is?' Covington said, when they were settled with their snifters. 'Does it have several organs packed into it, like lumps of clothing in a seabag?'

'That is a pretty idea,' nodded MacCracken. He was interested in the brain. He had dissected it into portions but found the process useless, philosophically speaking, leaving him always at the point he wished to start with—the mystery of being.

Covington sneered into his swirl of spirits. 'We must all have great heads or these qualities are small, trailing back on roots to be all fitted in like *turnips* or *yams*.'

'Another?' MacCracken held out the bottle.

Covington declined, saying the brandy was 'oily to his taste'. The comment irritated MacCracken extremely. It undercut his hospitality. Some friendships were better conducted through the mails, he swore to himself. In Covington's letters from 'Forest Oak' there was never any innuendo, while face to face he was full of it. As Covington placed his glass on a shelf and readied himself to leave, grunting and cracking his joints, his eyes were caught by an arrangement of shells MacCracken had placed in a window recess. Among them was a saucer of sea urchin spines. They were slim, shaded in brown, about half an inch long and dotted with small marks like goose bumps. Covington held them to the light like a diamond connoisseur, making judgemental clicks of the tongue, saying he 'owned these too,' having found them on the same island where MacCracken teased the tortoise and smothered the bird. His noise of disapproval was typical. He seemed captivated by the souvenirs, and somewhat lost, and yet there was this disdain also.

MacCracken cupped a hand to his mouth and shouted to be heard:

'Chatham in the Galapagos? When were you there? *When*, old stager?'

'When the world was young,' was all Covington would say in response. And perhaps that was everything, thought MacCracken, that he would ever need to say. (MacCracken would come to think so when he was wiser.) MacCracken was struck by the swirls of feeling that Covington shot out at him. They were full of pain.

Opening his diary that night MacCracken wrote his piece on friendship. 'No, a man does not have to be just as he seems. He can be more, in the light of understanding.'


Covington called another day, and they spoke of other things. Yet always with difficulty, one shouting himself hoarse, one scribbling figures, both barking replies. It was hard work, like playing a ball without bounce. It was all restricted to pounds, shillings and pence; to bullock wagons and weeks on the road; to sloops gone aground with their cargoes ruined and unwilling insurers. Covington carried a silver ear trumpet in a kidskin bag, but never used it. It was an advanced model, made to the latest design in London. He said he could have it manufactured in Sydney if he liked, and plenty would buy it, but had an aversion to forcing on others what was useless to himself.

'Such as your tale of woe?' MacCracken riposted.

'What's that you say?'

'Have you never killed a small bird with a hat?'

'Nay, not with a hat,' said Covington sourly. 'Not flippantly like you, *doctor*.'

'*What—is—it—about—you?*' MacCracken demanded, in a voice that made his brandy balloons ring.

'I am afraid to know,' replied Covington. There were tears in his eyes. MacCracken dropped his irritation, and embraced him as he farewelled him at the door. 'I have been a collector in my life,' said Covington. 'Birds and insects, small and large. Fossils. Mammals. Corals. You get so you forget what is man. You start to think, "Man? Why, he is just a stack of bones."'

'Dear Mr Covington, dear broken old heart,' said MacCracken. 'Tell me your tale. *Trust* me.'

Covington made a sound like a bull-seal smacking rubbery lips. 'Hmm?' Whether the plea reached him was irrelevant. Because the leaden door of deafness slammed shut. Because he tugged a lock of hair in ironical farewell. Because he trotted off into the dusk, a swirl of insects around his head, and gave those bothersome gnats all his attention.

Nurse Parkington urged on Covington a walking cure after giving him a good pummelling using pungent oils and the power of her mannish arms. Going about on his long, strong legs, Covington in a wide hat was like a patch of cloud-shadow on the headlands, trailing small boys who brought him bugs and rocks and other interesting finds. He could not stop himself peering, and MacCracken thought he was like someone hoping to find gold, he was so persistent in his hobby. What did he keep in his pockets? There was often a reek of raw spirits about him. He clinked and clanked like a bottle-oh. MacCracken saw him on the sandstone escarpment plucking at beetles. He saw him crossing the heath. He saw him up on the roadway, near the lighthouse, beating shrubbery with a stick and stirring up butterflies. He saw him coming down.

Steadiness and accumulation of effort defined Covington just as lightness of mind and quick snobberies defined MacCracken. Black boys threw pebbles on Covington's back to get his attention. Those narrowed eyes, all their shine burned out, turned upon miscreants and were calm the way a coral lagoon is calm in a ring of storm.

You may wonder why MacCracken thought Covington's collecting activities unremarkable in the man, even after his outburst at the door, and confronted with Covington's confessions still didn't know what he was talking about. The answer is that he ascribed them to fashion. Beetles were the wonder of the day in the Australian colonies through every class of immigrant. Stark wonder was the mood in the forest and in the house, with every piece of bark and every cookpot lid and plate left lying around lifted to reveal a creature never sighted before by civilised man and waving its feelers. There was a special pride among the takers of the place, because the plants and animals were so strange. Everything so queer and opposite. There must have been a separate act of Creation, it was maintained, and as Darwin had said on visiting there, to bring them into being. Swans were black. A mammal, the platypus, laid eggs, although nobody had ever seen one do it except the black fellows, who were not to be believed, so much of their lives being fanciful.

So it did not astonish MacCracken as much as it might have when Covington, holding a few struggling wrigglers in his outdoorsman's palm for MacCracken's admiration, gave their names in Latin. '*Leptosomus*, a weevil. *Ontiscus*, a seed bug. I was first to catch these,' Covington croaked triumphantly, 'in this very place where we are.'

MacCracken believed that Covington meant he was first between the two of them, referring to their mutual competitiveness. The doctor went on his way chuckling over Covington's clumsy pretensions. He was 'off' the fellow today. MacCracken had seen Covington's type at operatic concerts, men who had made their pile 'up country' clutching their programs as if they would strangle them, and popping their eyes from the effort of enjoyment and mouthing a few words of libretti taught them by daughters and wives. As with opera so with bugs. MacCracken put Covington in a box labelled 'Old Stager', smiled at him, shook his hand heartily, slapped him on the back, and made noises Covington would never hear.

Covington's other task on this stay was to attend to their business together. This suited MacCracken fine. After studying his ledger books Covington snarled, 'You need boxin' around the ears, young fella,' and took the books back to 'Coral Sands' and tidied them into columns. Then, without much ceremony, he was gone.

Thus MacCracken became Covington's beneficiary before he ever knew him at all. Covington's generosity fell upon the lanky Bostonian like spangles of light on the brow of a child. Covington's word was his bond in all his dealings, with everything down to the most niggling percentage point committed to memory. Not once did MacCracken question this generosity's foundations by asking himself what motive Covington might have beyond gratitude. As well question the loyalty of a dog when it came licking his hand, transferring its affections. The idea that Covington was looking for more meaning than a soul could bear and that MacCracken was the agent of that meaning would have struck him dumb.

'Forget the gold rushes and the delights of land-taking,' MacCracken confided to his friend Evans, the bookseller, as he placed an order for anything brand spanking new in natural history, 'if such friends as Mr Covington stumble into your days and bring you good fortune.'

After New Year Covington was back once more. They had their anniversary of meeting to celebrate with their eyes stinging in the month of smoke and cinders. Covington's visits became even more frequent, as curious to MacCracken as they were profitable. MacCracken brushed up his beetles and moths to compete with him, but had no chance of besting him—yet gained pleasure from the contest all the same. 'We have a good laugh and rub along,' he told his friends. Covington was an acute observer in entomology, just as in commerce and trade. MacCracken came away from their comparisons of beetles and wasps with a firm assessment of Covington's brainpower. When it came to birds he was incomparable, not just noting variations in plumage and beak-shape, but expounding anatomy as well. He knew the skull and breast-bones of skeletons by sight. MacCracken felt that whatever Covington turned himself to he was able to master, but at the same time felt a limitation, in that Covington was unwilling or unable to speculate from the foundation of the natural world into other realms of thinking.

'What is man?' was the old repeated question. 'What is life?' 'Where are these creatures from?' were others.

MacCracken wondered at the top of his voice, 'And what is their relation to the Great Flood of the Bible?'

'To the *what*?' The idea seemed to put Covington in a rage of stony deafness and ill-humour. 'What have pepper-pots to do with anything?'

'Did the Flood reach to all corners of the earth?' continued MacCracken. 'And if it did, look at the way these bugs float on water and climb to the tops of trees, and squat under rocks flat enough to stop their breathing but still emerge pretty fit, old Covington.'

'The sun is hot,' said Covington.

One day MacCracken showed Covington a seed bug, a leaf beetle, and a native bee—holding them cupped wriggling and buzzing in his pink palm—three random specimens that Covington stared at thunderstruck.

'What made you choose these—' he spluttered— '*Darwinii*?'

He reached out and dashed them from MacCracken's hand. MacCracken collected the same insects up, put them in an envelope and resolved at his next opportunity to have an expert tell him what in the world was offensive about them. Their discoverer's name? They had a pungent smell all together. He thought it was that. Deafness sharpens the remaining senses and drives an old walrus mad.

There came a day when Covington knocked on Mac-Cracken's door ready to say something, and—so it chanced —MacCracken had something more important on his mind and was embroiled in literary composition, and told his housemaid to have Covington call at a different hour, or even better, the next day, if it so pleased him and goodbye.

Covington turned instantly away. From that day the tone of the friendship changed. MacCracken began to feel its greater pull despite himself. Something more was expected of him and he knew not what. It made him grind his teeth in frustration.

On his next return, in April, Covington's mood of offence preceded him all the way up from Tathra to Sydney, and he declined to announce his arrival at all. When MacCracken went swimming at the end of his day's work he found Covington sitting in a rock-shelter smoking his pipe.

The next day, at the hour when they usually met, MacCracken found Covington engaged with the blacks of Watson's Bay, studying their fishing nets, employing their sign language (as they needs-must with a Covington), poking into their smoky fires with a stick to see what was sizzling there: oyster, mussel, pippi, scallop or abalone. In the dusk his great laugh told MacCracken his whereabouts. 'Do I care that you've found me?' that laugh implied. 'Rain on you, Dr MacCracken—I'll sleep under the stars.'

MacCracken peered into Covington's past as best he could. There were diffuse shapes down there like shadows in the tide. De Sousa, a shipping agent who was a rival to Smith and Elder, who handled Covington's cargoes, told

MacCracken that Covington had a patron who gave him his start in Australia by sending him there in the first place.

'A Spaniard, was it?' MacCracken asked. 'By the name of Sia Di?'

'That's no Spanish name I ever heard,' replied De Sousa.

'Nor I,' MacCracken reflected. 'But that is what he calls the man of his life. After which he gives a rueful grunt. There is a feeling of pride around the name, but Covington would rather spit than boast, so all I get is the annoyance.'

There was nothing unusual about obscure patrons in that colony. If you lacked one you invented one. A letter of introduction was all that was needed to change things around for a hopeful. There was no better place on the planet for correcting reversals of fortune and ill-birth, and none worse for sucking the spirit from an overreaching hopeful, either. Origins were made to be muddied as a matter of course. The next man you met could well be a lord, though he dressed in tatters and affected a colonial style.

So far Covington had told MacCracken nothing about his early life except that he spoke of a stepmother, a Mrs Hewtson of Mill Lane in Bedford. 'She had a loving affection for me, I believe.'

When MacCracken wanted more he got that lick of the lips, that beginning of words, that ducking away from straightforwardness:

'I am a man of no importance. Ain't that plain?'

The sarcasm was extreme.

MacCracken pictured Covington coming to his door the time of the offending rejoinder. He sensed the chance he missed. He could see the rectangular shadow looming, the hand raised to the knocker, the door opening, Covington's mouth opening, Covington's thunder beginning to rumble, and the words spoken by MacCracken's housemaid sounding—'Go away, come back later, come tomorrow, Doctor MacCracken is *thinkin*''—and then the jaw closing again, and the shadow creeping back, curled smaller than a beaten cur's and gone into silence again, to be held, to be held in hurt.

They carried on their negotiations in the open air, Covington with his hands hanging between his legs and a cowhide portfolio open on the ground. His knuckles grazed their receipts. Those disdainful lips were always half-smiling, even in a temper. That strong nose, with its powerful, flared nostrils, was a rudder to all Covington's mercantile instincts.

Percentage cargoes of cattle, red cedar, mutton fat and whale oil were the currency of the friendship in that time, before Darwin's *The Origin of Species* lifted a veil on MacCracken's understanding. They discussed the breeding of sheep and dogs, of which Covington knew plenty. He brought MacCracken a terrier pup named Spearmint, to replace brainless Carl who jumped from a cliff while chasing a kangaroo. He showed MacCracken how various defects in an earlier litter were improved in the terrier by a few simple expedients involving, said Covington, 'the wringing of small necks'. MacCracken gave Covington, as a gift, a shell collection he had made while crossing the Pacific. It made Covington smile.

To MacCracken's utmost surprise a day arrived when Covington bought 'Coral Sands' at the end of the row and came there to live. At this MacCracken discovered a Covington-like emotion knifing into him. Offence. He steamed and rolled his eyes. Covington?! If he ate raw potatoes and washed them down with earth MacCracken was ready to believe it. They were barely a village, a picturesque outpost of the great harbour. Yet they had their tone. 'To be attracted thither was to make a declaration of an artistic hue.' MacCracken's was a dalliance haughty in origin. Having scorned the Europe of his fellow-Bostonians he had found something superior. Call it his own Amalfi-like cove among the tumbled cliffs of an ancient land.

And was that Covington's motive, too—recuperation of ill-spent youth and convalescence of spirit? MacCracken thought not! He believed Covington a nobler creature— something like a horse. His move, MacCracken thought, had nothing to do with his own delicious ambience of face and manner. By his diary it was the twenty-fifth of March 1860. MacCracken itched around the house and ranted into his breakfast (fried schnapper, fresh rolls and newly roasted coffee). Did he want Covington as an acolyte? Was Covington in love with a doctor as with an all-knowing God? What about Mrs Covington and the children

MacCracken had never met? Were they to fend for themselves in the black starry nights of Pambula, where tribesmen built fires in hollow trees, and danced their corroborees clicking spears as Covington had told him, within sight of the substantial residence (with attic windows) named 'Forest Oak'? It was where Covington wanted to be buried, on a nearby headland, but, 'Not too soon, MacCracken, if you don't mind.' What about the corn flats and the seven hundred head of horned cattle being driven down from a high plateau to those grey box-log wharves that Mr Covington had built himself? Half MacCracken's cattle they were! His risk, too! And what about the high plateau itself, where the Covingtons had a farm they all loved, and where the Covington children, older boys and younger girls, lived a natural life of horse-galloping and calf-chasing, and despised the idea of 'town' and their father's 'retreat'? Mrs Covington apparently felt the same and hated to leave her brooding hens, her milking cow and her pet cockatoo.

All Covington's mysteries had a special interest for MacCracken. But he wanted them unravelled in their rightful place, in those 'up country' locations, as Covington said (where up meant down). MacCracken had contracted Covington's corn to a miller. He had paid for half Covington's cattle in promissory notes. He had expected Covington, though he never said a word of it, to supervise their business on the spot. He had his picture of Covington at his cattle yards, dressing him in imagination in an oilskin coat, wielding a fly swat and puffing a corncob pipe. This was before MacCracken ever saw him wearing an Argentine poncho and smoking thin cigars, thus confounding his every last supposition about him—that he was devoid of romance absolutely.

The first night of Covington's new residence there came a banging on MacCracken's door. An imperious, troubled

Covington stood there, a fishbone stuck between his teeth that he wrenched free while bellowing:

'MacCracken!'

'I am here,' MacCracken stabbed his own chest (to the deaf 'un).

'I want your gun. The ladies' shotgun you keep for scarin' possums off your roses ... '

'If you want something,' said MacCracken, opening the broom cupboard where he kept the slim weapon, 'you must get out of the habit of disparaging me for having it.'

He slapped the gun into Covington's hands, who examined it curiously, underlining his ingratitude. 'I picked you for a breech-loading man, MacCracken. Such a convenient, slick conception of shooting they are. I meself am wedded to the old frontstoker. What are these charges made from, fly paper? They are somewhat clingy to the touch.'

MacCracken only stared at the phenomenon before him. A gentleman without pretension when it came to securing a friendship—who made scorn his starting point. It was quite admirable, really—if you liked that brand of swank.

All this was by way of an aside, however, because Covington then said:

'MacCracken, answer me this ... '

MacCracken only half-listened, and surreptitiously consulted his clock, expecting Miss X and her nightly visitation quite soon, not wanting Covington to stir gossip before he must.

'I won't *keep* you, MacCracken,' he sneered, 'but where is God?'

That caught MacCracken. *God?* Covington waved the gun around and asked MacCracken questions from his experience as a surgeon. 'In your dissections at the Boston hospital, and ever after,' he wanted to know, 'have you seen evidence of a human soul?'

The questions rained down on MacCracken's slightly bowed head. Had he cut a man open and found any immortal part? When he was in Covington himself did he peer around corners? (No.) Had he ever cleaned out a possum? (No, he had never cleaned out a possum.) Had he ever cleaned out a bird? (No, he had never cleaned out a bird.) If he had, would he ever have found any difference? Was he not an educated man, compared with Covington, a dunderhead? So? And so on?

'The difference, MacCracken, *what is it?*'

Difference between what and what? (MacCracken mimed.)

'Difference between a man and a rat, noddy-foozle!' Covington bellowed. Then grabbing MacCracken by the elbow, he dragged him into the night. They stumbled along the shelled pathway leading to Covington's gleaming, newly painted and snug-as-a-ship's-cabin cottage with its lamplit nameplate 'Coral Sands'. Within that whitewashed palace, through a square window-pane, MacCracken saw a short, humorously squat woman in voluminous skirts standing on a seachest and flapping her arms against her sides. She had jet-black hair pinned with a half-dislodged Spanish comb.

'Shoo!' Her eyes bugged and MacCracken saw why. There on the dining table, consuming an end of loaf, was a sleek whiskery rat. 'Get it off!' she shrilled.

It was at this moment MacCracken realised that every word spoken to him that night by his friend, from Covington's first banging on his door until now, had been bitterly sarcastic in tone. The sarcasm was not addressed to MacCracken. It rose to the height of the universe itself. Within that dome, rampant as a flea, stood the Mrs Covington that MacCracken had thought in their country fastness. He found himself commandeered as an illustration in a domestic tableau: 'MacCracken will show you!'—and realised he was no more real to Covington in his rage than the man's own face in a mirror.

'Dr MacCracken,' Covington silenced the woman with a roar, 'says there *is* no God! And so bless you and shut yourself up, darlin'. A man is no different from a rat, and a man never frightened you, did he, eh?'

'Blaspheme all you like and still say your prayers. Heaven prepare you, Syms Covington,' she said.

Covington waved a finger under his dame's nose, and while MacCracken stood feigning amusement, he raised the ladies' shotgun and with quick aim sent the rat slamming against the opposite wall without breaking a dish.

'If you'd done that first,' said the wife, hopping down, 'we'd a-had none of this nonsense.' She dusted her hands and introduced herself, giving the doctor's hand a vigorous shake. 'Mrs Covington, and very pleased to meet you, sir. I'll put the kettle on.' That done, she skipped back to MacCracken's side while Covington stood nearby cracking his knuckles and staring into the night. 'He gets his ideas,' she lowered her voice and turned away from Covington. 'I do my best, but Dr MacCracken,' she drew breath, raising herself to her full height, '*he* has such hopes for you!'

MacCracken allowed himself to be soothed by a cup of the best China tea. Then, with a neat bow, he bade them goodnight and went to join his pleasure.

# BOOK

## 2

*On a Thousand Gallons of Blood*

# 1829

They weathered howling destruction in the Bay of Biscay, making course for Lisbon. Fear could be read in sailors' faces but the one face to watch was John Phipps's, that never showed anything but faith in his God's intentions.

Covington puked and moaned. Watched England sink like a plate in the suds. Felt he was being stolen but the feeling did not last. With stomach cramps doubling him over he learned to scramble and fetch. With a bar of holystone he scuffed decks hairier than any storms, and whenever he faltered looked around for John Phipps, his begetter in the life of the sea, and was rewarded by the sight of him, who was always in the act of obeying an order, running along deck in bare feet, hauling on sheets or hoisting himself up into the ratlines, his dark narrow face raised to the sky and seawater running down his beard.

When Covington was over his first spewing he knew he would be strong. He was hungry and chewed on a heel of bread and a rind of damp cheese. When the sea rose in a green wall, striped with foam, it amazed him and he started singing. He and his small friend Joey Middleton, who was in the same mood, practised sliding between-decks. Their bark, the *South Sea Castle*, flicked about like a fly in a bottle, and the worse it got the more they hooted and hurrahed, calling on the heavens to do their utmost. They were like infants in a playpen the way they carried on. It made John Phipps somewhat proud. His band of pilgrims were all ship's boys now, nippers, odd-jobbers, and officers' servants. They spread their joy and it was a happy vessel. It was good to hear Phipps's accents of the countryside when the watch was changed at four in the morning, and the boys were called to vacate their hammocks:

'Oi, sleep any longer and ye all shall be hanged. Come on rascals, Slow-pace Dell, out of there Foul-wind Wingate and Crook, tip you over Sleepy-head Middleton, will you?' When Phipps came to Covington he prodded him on the buttocks. 'Is it the girl called Dull?'

'Aye, and if you keep it up I will kiss you.'

When they were some way out from England their Christian captain faced his duty and called for the cat-o-nine-tails to punish shoregoing offenders. A few had rioted in Portsmouth and delayed the sailing.

It was fearful to watch. The bosun's mate braided a whip from rope and quarter-inch line, and fitted a red baize cover to the handle. The boys saw drunkards and malingerers baring their backs and taking their punishment with a rag between their teeth. 'Here is an example for you, a foretaste of hell,' said Phipps. Bare human skin was a delicate parchment exposed to an underlining of blood and lymph by the dozen strokes the bosun's mate made. The captain and the surgeon watched as the cat was applied with pernickety precision. The very first lick caused a man to jerk his body around, while on the last he slumped. The sight made all the boys sick and they swore it would never be them. Covington stared in pity. Animals herded by his Pa in the Bedford stockyards were never flayed as fiercely, because their hides were differently valued.

T he great trust that Covington had in the world's advancement of his fate, that he was born to and found only rarely shaken, he brought with him from Bedfordshire to the sea. It only rarely rankled that his duties were all odd jobs, and that his cheerfulness was taken advantage of in the same way his big brothers had used him. He swabbed decks, nailed boards, mended barrels, carried buckets of food to the officers' wardroom— and longed to be chosen for the clerk's job of transcribing the ship's log, so he could show his full mettle—but, whenever a messy task was to hand, such as plucking a fowl or despatching a goat, his butchering sire was invoked, and Covington was called upon to wield the knife.

When he came up for air he struggled forward along the spray-whipped deck and clung to the bowsprit. It was his instinct to dive into water to clean blood from his arms. The whim had to be fought aside, even on a ship sailing the depths of the ocean. Covington dangled over the lively green swirls watching the ship's figurehead dip and dive. She was a maid with hair streaming back, with small bold breasts and although she was just a piece of wood painted in simple colours, he fell in love with her, imagining himself to sleep at nights stroking her. She had fluted nostrils, dozy eyes, and a wide strong forehead.

A feeling Covington had in dreams of celestial highways

and lightbeams in clouds was translated into his experience of naval voyaging. Later there would be ghostly passages through narrows, blue fire trickling around masts in electrical disturbances; and there would be a giant ray flapping under the ship—it would be as big as the ship itself—and when the undersea animal turned and came beating its way back towards them as if to engulf all their timbers in its maw, Covington would have to be restrained from jumping overboard because of a wild conviction that got hold of him that he should wrestle with the creature and save them all. It would be his third ship by then.

The young man climbing the stile—John Bunyan's Christian rendered in coloured glass-panes—appeared at sea one day when Phipps threw Covington a rope as he staggered along the deck, and saved him from spinning overboard.

'Thank Christ, John, it was you,' the boy spluttered, his hero-worship at its peak.

'You thanked the right one first,' Phipps confided in him: 'As for me, had I a thousand gallons of blood in my body, I could spill it all for the sake of the Lord Jesus.'

That night when Covington stretched himself to sleep he pictured Phipps wielding his sword repelling enemies, and the great admiral, the one that John knew, coming across the water on a shining barge to see all was well. This was the way Covington stayed safe on his path—his adventure out from home. John Phipps had Christ and Covington had John Phipps understanding Christ, and that was the truth of the matter.

Hammocks were piped down and it was sleep, palms held together under their chins in an attitude of trust. The sails leaned them into the night, high-walled and straining as the helmsman brought the vessel constantly up to the weather. Trickling and sloshing sounds accompanied them the whole way. They heard the vessel's timbers groaning apart, admitting water, creaking shut.

There was always a bias of gravity to one side or the other of a ship, a slope running up or down determining which way the boys went scuttling, with tin pots and lanterns clanking, and loose items sliding across the under-deck to tumble around their ears. There was always a dampness, too. Below decks it was dark like the lock gates where Covington used to swim. The oak planking was always wet and looming, threatening to split. Yet it was strange, considering the violence of the sea: there was always a feeling of safety, too. Everything could be trusted to have a good end.

One day when Covington took his turn with Phipps at the bilge-pump, and Phipps asked him why he grinned when the work was so tedious, Covington said, 'I am always full of good notions that come into my mind to comfort me as I pump.'

'You have a great innocence,' said Phipps. 'Be sure that you don't get taken for a fool.'

Covington laughed at this, and stamped on Phipps's toe.

Phipps saw to it that they messed together, all five boys in a bunch, and saw to it that the one-armed cook never held back on the duff, the sea pie or the lobscouse, nor pease soup or even grog, much watered, when his boys scrambled for their dinner. They called at Madeira for victualling and ate grapes for the first time, also figs, melons and dates. They kept oranges in nets hung thick above their heads, and had a rule of honour that they were to eat one only when it started to go bad. The other two mess-mates were Door and MacCurdy, foretopmen with John Phipps. They were great simple West Country fellows and like big brothers to the boys. While Door carved the meat MacCurdy wore a blindfold and called out names, so that nobody could say they were given short rations through malice, or were given too much through being favourites. When excitement mounted Joey Middleton danced on their mess table, ducking and weaving to avoid the lantern and overhead beams—then went crouching with his back bent like a frog, as hands reached out and grabbed him by the skinny ankles.

When they came into the hot latitudes, Phipps taught them how to make palm leaf hats, as favoured in the Caribbean islands. In talk of those islands Joey Middleton's hopes of finding his father, the red-headed sailorman, were dashed, because the *South Sea Castle* made her course in a contrary direction, past Brazil on a survey of ports eversouthwards. Covington learned carpentry and odd jobs in that first season away, and learned to play the fiddle with seaman Door his teacher—long afterwards being able to recapture his days afloat from a sniff of tar and a shaving of timber, and from the quick bounce of a well-powdered bow on catgut.

They were starting to be sailors and Covington felt pride in his new craft. As their Sea-Daddy John Phipps oversaw the boys' cleanliness and their clothes, teaching them to cut out jackets, shirts and sailcloth trousers, and to sew, and to wash it all in a sea-bucket, and to mend their clothes as neatly as apprentice tailors might.

Phipps was a great one for avoiding any show of favourites, although Covington had a vanity that he was the one. If he played 'To Be A Pilgrim' in a key of G on his fiddle, low as a bullfrog growling, it brought tears to Phipps's eyes. Their special closeness was Covington's secret and at first made him proud, then careless, then itchy at the restriction. For he grew past the conceit Phipps had that they were all his honest pilgrims. Covington had too much blood in his body: it pounded him along.

P art of their orders as naval surveyors was the getting of creatures. On a wooded island a party went ashore with guns and ropes. John Phipps was well known as a countryman and favoured by the officers for the hunt. Phipps named Covington his offsider. They chased pigs with the help of a slavekeeper's dogs and cornered a great tusker. Covington went in with the knife while ropes held the animal down. It took six men to carry the boar back to the beach, where it was divided into portions and rowed to the ship. There were fights over the crackling after the cooks had done their best, and the bones were cleaned for the surgeon to measure and compare. This wooded island off Brazil was the first place where Covington ever ate a banana. It prickled the back of his throat and gave him a rash on his arm. It was the first place he ever saw a slave, a tall man and black as a moonless coalheap, with skin of sweating lacquer as he ran with the dogs on long swift legs. Phipps fumed at the practice of slavery, and gave them his earful about it: 'See how he's bred for the chase.' But his indignation wore thin with the boys when they saw the slave whooping with joy in the hunt. He was in there for the kill, and as one of the officers said, 'Went in as well as ever I saw a man in my life.' Besides, Covington saw the hut where the man lived. It was all done up comfortably.

'Ignorance is your name,' said Phipps to Covington, 'and as your name is, so are you.'

Covington whistled and jumped in the longboat, standing in the bow with his bloodstained arms folded as the oarsmen rowed them back to their vessel. He found he could give John Phipps the freeze, and make it last for days.

Though they broke that coolness soon enough, when Covington saw Phipps hanging upside down from a hatch and laughing in his face. It was in a storm.

'Are you afraid, Syms Covington?'

'Not if you aren't afraid.'

'And why am I not afraid?'

'Because of the admiral who sails with you, John.'

'And who is he?'

'He is a revelation of Christ to our souls indeed.'

Phipps made an action of kissing his hand.


One day they hooked a great shark and brought it smacking up onto the deck. The boys were ordered to do their worst. They sat on it straddled in a row, singing a shanty, cutting chunks from its flesh while it was still kicking. Most of their days they lived around the mess table, in knots of alliance or in brotherly rivalry, buzzing to get the approval and admiration of the older men, and then spending time imitating those older men in their swagger and being cuffed around the ears for it. They were all on top of each other in the vessel as Covington was with Mrs Hewtson's sprogs. Was there never any leaving home in this world?

Because their survey ship zigzagged around and retraced her steps taking chartings, other ships from England passed them and letters from Mrs Hewtson and packets of plum-cake wrapped in muslin awaited them at several ports of call. Though Covington was generous in all ways, this cake he kept to himself, nibbling at mouldy portions with his elbows tucked to his sides, his wispy-whiskered chin covered by a hand to hide his actions.

During their time at Rio he made a good bargain, and bought a dancing master's fiddle that fitted in the pocket of his seaman's jacket. They said he would play tunes into heaven—he had the knack. 'In the book of Revelation the saved are compared to a company of musicians that play upon their trumpets and harps and sing their songs before the throne.'

It was at Rio that Wingate, Crook and Dell left the ship to join another bound for England. Now in that secret society of prayer in the Bedford style there was just Covington, Joey Middleton, John Phipps and the sailmaker, Mr Harper, who wrestled with his conscience all the time and was a serious devotional man. Then they, too, changed ship, the close bunch of them along with Seamen Door and MacCurdy, taking berths on the survey vessel *Adventure*, under Captain Phillip Parker King, a colonial-born officer of modest temperament and good navigating skills. They were to make their way farther south.

King had his son aboard, Phil, who was a midshipman and the same age as Covington—thirteen, now—and so they were like a family of scrappy brothers all in their various cramped messes, boys doing their growing and their bullying and their mauling of each other as they bobbed around muddy river mouths learning to be tars. King was boastful and tricky to play with, being the captain's son, having a spirit that checked itself when others wanted any game to continue. The point came in their japes when King fell silent and dropped his easiness, lowering his chin and clenching his knuckles white: 'I say you fellows enough is enough, and I really think we shall lose our goat overboard if we set her loose, and you, Covington, get the bucket off her tail.'

'I didn't tie it on.' Nor would he ever, either, knowingly

taunt a dumb beast. '*I* didn't tie it,' he repeated, levelling his gaze at King.

'Quick now, or *it* shall get you a flogging.'

It was King who'd slipped the knot, giggling with excitement, his sallow eyes flicking around the faces of the gathered boys in his haste to get their approval, which he then so snivellingly traded away.

After Rio it was the wide-mouthed La Plata and Monte Video, where they anchored so often that the ship's company thought of its waters as an English possession. On still afternoons, drifting across the water, there was always the smell of roasting meats coming from the cookfires of the poor. Sometimes following their afternoon drill the captain gave orders to dance and skylark, and Covington led the whole barefoot crew on his miniature fiddle, standing on a barrel while they jigged and sweated with their hands on their hips and their heels kicked up.

*Brace up the yards and put about*
*Cut a fine feather and fly*
*Give her a foot, she'll go like a witch,*
*Sail till the seas run dry.*

It was his and Joey's emblem song and they danced till their toes grew blisters. Covington named his miniature fiddle Polly Pochette and studied sailors' ditties, and for the entertainment of the officers played 'A View To A Kill' and 'Old Greensleeves'. When he gave out with the 'Twenty-third Psalm' it silenced all, both saints and sinners. It made them understand they were on a journey, and it was not all to do with charts and chronometers. There was another way of measuring man and other beacons in the dark.

A whole two seasons passed in that walnut shell, man-oeuvring around weedy river-mouths and making their survey with particular care. 'There was nothing prouder afloat nor on the land neither,' it was said. The *Adventure* was the world for them. If a bird sat on their rail or swayed in the wake with its wide wingspan hardly moving for an hour, then it was a special creature blest of an understanding of where the peak of life was. If they saw the same bird in another place, on the shore, say, or on another ship, it was nothing—just a bunch of croaking feathers. Phipps's pilgrims grew past his influence but were still attached to him—more than he realised but less than he desired. Each officer, each rating, each boy—they looked on each other in the way they looked at the bird, and there was an aura of belonging about them. Even those who came up for a flog-ging felt that. They knew that for thieving a shoe or cracking a mate's head ashore they would get prison or life transportation; whereas on the ship a flaying of the corp-oreal body was all they got.

Covington learned his Spanish on a seawall by the muddy La Plata. He played ball ashore, using a dusty bladder. But mostly the box of planks that was their ship was all the room he needed. When the boys heard the bosun's whistle and came dashing back to the dockside, and saw their ship anchored out in the bay with all her sails furled like bleached cigars, they were glad in the way of someone in love.

Their superior pal Midshipman King was born at Parramatta and came to England as a child. He lazed on the booms and told tales sillier than any topmast spinning. He told of Australia and the great hunts that were held there, horses leaping fallen logs and chasing kangaroos that were swifter than any deer. Covington heard about the stubborn wombat, with its baby the size of a pink fingertip clinging to the fur, and the ant-eater so covered in spikes that if you wanted to nurse one you would have to wear armour. King told of the duck-billed platypus that was fur-covered, had poisonous talons, and laid eggs in the banks of streams. It was part of the *Adventure*'s commission to collect samples of natural history, and King disappeared sometimes, in the company of his father, bringing back buckets of shells and the carcasses of animals which they bade Covington skin, which he did willingly enough, as his father had shown him.

The boys wrestled and played cat's cradles. They were adept at knucklebones on a heaving quarterdeck. They confided their dreams. Covington thought King's dreams were fine—he wanted nothing more than his father had—though if Covington had wanted just what *his* father had, then he would wallow in nothing but blood and be condemned forever to goat-slaughtering in their ark. He wanted better, and one day the chance came. With Volunteer Musters, who was eleven years old and had lately joined their ship, King, and Joey Middleton making a fourth, Covington was chosen for grammar and penmanship lessons together in the poop cabin, under King's Pa.

Covington, being the best schooled of all the ordinary boys—amazing the captain with his flourishing y's and curvaceous p's—now had a dream that he might be selected from the ranks and be made a midshipman one day too. He was in no wise duller than Phil King except in mathematics. So was he vain?

John Phipps narrowed his eyes on Covington when he came spilling from the schoolroom, excused other duties for a swot. Here was the lad he fancied sometimes to be mistaken for his son, in the lunacy of his desires—a son to go round anywhere in the world with him. And so he thought: *May he never lose his soul to advancement.*

Covington approached his childhood's end when they shipped south from Monte Video in the southern spring. Farther and farther from home, and then Phipps, once so easy with his lads, started to fret that he would lose this one altogether. Covington's voice breaking was like gravel and he had reddish-gold scruffy hair sprouting from his underarms and crotch. He was heading up for a fight with his mentor, and Phipps cursed that he knew it from the start, from the day Covington had tugged himself with unmistakeable wantonness in the green canal water and called, *I caught a fish, it's a big 'un, look, see?*

It emerged that John Phipps, with his special mission in Christ as expressed through his young male charges, was a calm controlled man only to a point. For he was made for anger like anyone else blocked from their wants. He wished, instead of the cheery laugh he'd given that time when he first clapped eyes on Covington, that he'd leapt in the water and given the naked boy a hiding.

Meantime Covington's cheerfulness could be heard booming through thick timbers, even over the roar of the sea. And Phipps said: 'Hold you there still, who is that?'— as if he didn't know him.

Another time he said: 'A fool is someone that having the liberty to keep what he has, loses it anyway.'

To which Door replied, with a wink at MacCurdy: 'Then

you must be a big fool, John Phipps, because you dream of nothing else with your hands on your cods, and toss yourself overboard every night.'

'You see how the bee lieth still all winter,' sneered Phipps, 'and bestirs her only when she can have profit and pleasure.'

'I remember what I wuz at Covington's age,' said MacCurdy, 'and am no different now.'

'What wuz you, shipmate?' asked Door.

'Full of fuck and half starved.'

When they turned around they saw that Phipps was gone out of their hearing, leaving his tin basin upended on the barrel where he sat.


Every day the weather became a few degrees colder. After three weeks they came through misty passages around Cape Horn and there was snow on the decks. They hove to in dark narrow places, crags to each side of them and wild seas streaming past. The ship hardly rocked at all, but was held in a clammy, nerve-racking stillness. Covington's cheerfulness was a beacon in that weird southern light— even as word came that one of the survey captains, Pringle Stokes of their sister ship, the *Beagle*, had gone mad and stiffed himself in the wild nowheres. He had used a gun on himself and command of his brig was given to the aristocrat, FitzRoy.

In Tierra del Fuego, land of fire and sleet, Covington showed off and went bare-chested in the blistering cold, bellowing like a bull-calf every chance he got, sending other crew members into fits and encouraging obscene speculation. Meantime John Phipps's thoughts burst like red hot boils.

Phipps said Covington was mad, but the rest knew it was not that, but was only a boy's build-up of juices turning him round.

Phipps felt personally accused all the same. 'Come to our prayers,' he whispered, and got all the boys together. Covington attended full sweet and dropped to his knees, being so good-willed about everything—yet Phipps while mouthing his worship could see Covington's eyes flicking around with humour, picking holes in the coats of the godly. Phipps threatened him with hellfire but the instinct of self-preservation was not in the boy at all.

'I doubt that God hates me and you said so yourself, that God loves me for who I am.'

'All praise to you for remembering so conveniently,' said Phipps. 'Anything can be argued from the scriptures if self-will be a guide.'

'I am not arguing from scriptures, I am arguing from you.'

Phipps blamed Covington for associating with carnal, loose, and wanton shipmates—the same pair he had got his boys messing with. But he now said that Door and MacCurdy had deceived him with friendship. The two big men, in reply, told Phipps he was a vain fool, and on the first Sunday of the month they arranged to be moved to another mess.

'Will you go with them?' Phipps asked Covington.

'Why should I?'

'To have your laughs.'

'I will have my laughs wherever. Jesus was a boy, and so he must have grown to be a man in the same way I have—'

'What is that way?'

'Shall I tell you?'

'I am asking.'

'You will condemn me for saying.'

'I promise you I shan't.'

'With his *prong* sticking up in the morning.'

'Blasphemer.'

'This is so like you, John.'

'It is your aim to wound me.'

'You never said there was anything wrong in me before. Nay, you praised me, and said I had no feeling for sin, and so I wasn't a sinner at all. But now you say I am, for the same reason.'

'It is what you do next that overshadows you,' said Phipps.

Covington smacked his own head. 'You are trying to drive me mad.'


There came the day on their survey of those coasts when they saw women of Patagonia swimming for shellfish in a wild surf, emerging dangled with kelp. Covington saw a young maid among them, tender to be observed, nakedly diving. She had black hair capped wetly. They were on a shore party when the lads bawled, 'Take 'em while they're hot,' and Covington it was who won wagers by wading in and grabbing hold of one.

The lads hooted: 'Covington is on his way to Hairy-fordshire to play at lift-leg,' and then they bawled: 'He owes the bull shilling'—money paid out for drinking after being seen with a woman the first time.

He had her in his grip for a second. She was like holding a slippery fish. Her eyes were brown and shone with water droplets. Very wide in the forehead and curious about him too. She did not smile, because smiling was not their way: but she enjoyed his play, he believed, even in her savage fear. Then she was gone, running along the shore, leaping oyster beds and jumping into one of their canoes that always had fires burning in them.

The captain hailed the canoes back, and got a few of their men up on the deck. He bade Covington be clerk and write down some of their tick-tock sounding words: he got the ones for duckling, feather, fish and a small fly, before the subjects lost interest. The surgeon measured their limbs. These natives, he declared, were built like whales or dolphins, all smoothness with rings of fat on their abdomens. The surgeon made comparisons with the muscles of sailors. He turned to Covington and said, 'See, the ship's boy is built like a smith or a porter, the form and size of each muscle may be traced in action,' and asked Covington to lift a block and tackle, traced a pencil down his arms, and then asked one of the natives to copy him, and the fellow, with a laugh, hurled the block and tackle overboard. The crew milled around with all the excitement of a fairground show of freaks and the natives all dived overboard.

Covington stepped back, and turned—finding himself faced with John Phipps, who let out a stream of invective, and said *she* was the very breathing devil. Phipps checked that no-one was looking and then he slammed Covington across the shoulders, holding him face-in against the side of the poop cabin, his nose somewhat crushed, and told him to leave well alone.

Shivering sick for the sight again, Covington tasted rebellion and blood from cutting his lip when Phipps pushed him so hard. Who was she that rose in a vision with blurred eyes dripping with diamonds, and stopped his heart? Previously it had been the young man leaping the stile. Now it was a maid who dwelled inside his head and fired his thoughts. It was part of his pride in his struggle with John Phipps to keep in his mind what was best in his mind. Thereafter Covington believed a maid would rise from the water some time, as it were riding in a seashell in wild-born finery and savage guile, and join his dream of what it meant to be alive. He bedecked her with beauty in imagination. He was fourteen years of age now, and as good as any man. He crouched over her and she stared up at him and asked him his will. Meantime the officers all came out and stood on the quarterdeck as the natives cursed the ship and the canoes made a pass, this time with shouted invective and plain threats—and the officers primed the guns but said it would be shocking to fire on such poor miserable creatures, as they might otherwise have done. Then they sailed away.

The next time their mess sat in a circle, with Phipps giving them their catechism as they mended their clothes, Covington was present among them, as innocent as if nothing had ever happened, and Phipps asked if he was back in their fold.

'You see where I am,' he said. 'In my place, on my barrel, with my spoon and my bowl.' They held each other's gaze, and Phipps was the first to break it.

Meantime the lads kept at it: 'Covington is randy as a dog, a proper Spunk Tub,' and teased him with tales of shoregoing delight.


They landed at the island of St Helena on their homeward voyage. Door and MacCurdy, who were heartily sick of their righteous friend, got Covington three steps ahead of John Phipps's fury at what was going to happen next.

Away past Napoleon's Tomb they bribed a black sweetheart to take the boy for an amble upon the hills, with instructions to give him his ro-ho-ho if she would be so kind. She had broad hips, short legs, a hooked nose and hooded eyes. It was the eyes that caught Syms Covington, looking out like cats' eyes from a cave. Her laugh was low and rough as combed honey. Her name was Hickory and she was promised in marriage to an old man. Meantime she had her freedom, and it came to the forefront of the boy's mind that what was freely owned was freely done, and if not, then wasn't it said that doing it with a black woman counted for less? They peeled and ate sticks of sugar cane, which she jerked in a frank rude sign, and she drank from a spring, putting her rump in the air, giving him a look at her. She giggled and rolled her eyes, making him laugh. She showed him a secret pathway back to the town, with a knoll of prickly grass where they sat together. She startled him, pulling at his clothes. 'Now we play hunt the dummy,' he thought to himself. Out flipped his Nimrod from his trousers, and she took it in her fingers, trilling '*Olay*'. She favoured him, then, in such a surprising *movimiento* that he forthwith spilled his milk.

It made John Phipps surly seeing Covington return whistling to the longboat, tossing his pigtail around that 'the foul maid Hickory' had loosened in her play.

'They have made a fool of you,' Phipps said. 'Did you see her go with the next who waited?'

Covington tried telling him of his pleasure but Phipps spat sideways. He said the moll was a Madam Bubble, a witch and a mulatto mutt's true bitch to fiddle a fiddler so. She was the way of the world and the evil standard of going ashore. 'I know this about you. You care neither for man, nor argument, nor for example. What your mind prompts you to do, that you do, and nothing will change you in your way.'

'I own freely what I done,' said Covington. 'Remember that time—when you thought I was the easiest soul in all creation, not even knowing who I was?'

'I do.'

'So what of it, John?'

'"What of it?" It means you are without conscience.'

Phipps took him by the throat, bringing tears to his eyes, and asked him his meaning towards *him*.

'Nothing towards *you*.'

'Ah, I am nought to you?'

'That is not my meaning.'

'I think it is.'

'Very well. "Nought to you,"' the boy replied, pushing himself off. 'If it is what you say. But I would never say so.'

It registered on Covington's brains that he had grown half a head taller than most of the crew, that his fiddler's wrists were not so much green and supple any more, but strong as weathered spars. He supposed a reckoning with God would come for his carnal sins, for all his catechising said so, but he took a chance on that and trusted his heart. As they sailed north for England, re-crossing the equator, Covington celebrated his fifteenth birthday with a boast about fighting prowess, and was subjected to hot wrestling on the foredeck, with John Phipps the ringleader against him in testing his mettle. Phipps scored a bloody nose. Seaman Door cried mercy from his handsome face, and two favoured boys, Midshipman King and Volunteer Musters, stood by smirking proud, for they fancied themselves connoisseurs of the ring. 'You shall be our hammerman, bedad,' said Musters, adding: 'Covington *my man*.'

Covington thought this pretty chip of gentleman Musters, who was aged eleven to his fifteen years. There was now a difference showing between King, Musters and Covington that made him smart with secret tears—a matter of who was born to advance. Would it never be Covington, with a head some said was handsome but others taunted was like a mangold-wurzel on a pole, who spoke in the accents of horse-markets and addressed his geometry and algebra aloud, as if they were living creatures? Who consorted with the gentry of the ship less often now that he was no longer a pup, and padded between decks carrying buckets of steamed cabbage and green potatoes with a look of stunned devotion on his face? Who because he was marked to slit animals' throats tended to them more considerately than anyone else, and so smelled like an animal too?

Thus Covington's freedom of the ship narrowed, and on their last leg home the *Adventure* was a smaller, more crowded vessel in his brain than it was when he first stepped on her decks. He spent more time on his own, wedged in his favourite place in the bows considering the innumerable waves and allowing the odd one to rise and smack him full-force on the chest. Joey Middleton remained his one true mess-mate. Joey's growth had not started yet; he was a splinter compared with Covington's tree-trunk, and needed defending from bullies and flirts.

Every act of kindness shown to Covington he passed along back to Joey, and had it returned threefold. When Covington was sad, who sat at his feet? When Covington went dancing wild, fiddling and making practical jokes around the ship—taking the captain's turbot from the fish tank and having it served up to the midshipmen—who always tried to outshine him? When Covington left a space empty in the affections of John Phipps, and Phipps refused to allow him to fill it any try he made, because of a *pride* in Covington and a *convenience* in his morality, who was it that sidled up to Phipps and took him by the hand, and tried to persuade him otherwise? On all counts it was Joey, pale as waterweed, attentive as a barnacle, always behind in making headway with sacred texts.

Sometimes Covington and John Phipps caught each other's eye, and in the hustle to be about their duties spared a moment for each other. Then it was all the same to them as if nothing had ever changed in their affections.

'It's a wild night we're having, Syms Covington.'

'It could not be wilder, brother.'

'Only a few more watches and we'll see the Scilly Light again.'

'Here's hoping we shall not make feed for the fish.'

'That's a dismal thought to have so close to home.'

'Anyway it's my prayer.'

'Oh, your prayer. Mine is to make worship together on dry land if we can, and not be about breaking each other's noses.'

'I had the same thought, John, but where did it take me?'

'Come down from where you are.'

'Look for me in Bedfordshire,' said Covington, 'or not at all.'

'What, like a needle in a haystack?' taunted Phipps. But they both had a mind-picture of a rural barn where they first gathered, and a chestnut tree in a shallow vale where they had bowed their heads thinking who they were, whence they came, what they had done, and to what their King had called them. It was all of England to them, that place. Would they meet there, ever, and mend their breach?

Covington returned to the city of Bedford a sea-hardened young man with elbows sharp in crowds and giving out a feeling of experience bigger than he knew. Everything about his birthplace looked shrunken in his eye. The Quentins' hide-house where he had worked was laughable, foul, the distraught faces of the clerks despicable. Likewise the pretensions of the owners, who had erected a plaque on the front wall of their works celebrating the construction of a sluice that led from the tanning vats to a piece of boggy vacant land which reeked like the cesspits of Beelzebub.

His pretty stepmother—who in his mind's affections he had sometimes bounced on his knee in anticipation of this moment, and kissed with surpassing warmth—had ballooned in size. Mrs Hewtson presented a chubby, breathless version of her former self when he banged the knocker of the house in Mill Lane. She filled the doorframe, almost fainting in surprise, and displayed a deference that somehow offended him.

'It's our boy, I wouldn't a-known him,' she dropped her chin.

He felt a stranger as he sat on a stool with a teacup in one hand, a slice of bread and dripping in the other, making small talk in the room that had filled his heart with comfort just by thinking about it on cold and stormy nights. His speech was peppered with sailor's slang, and if he said 'brace up', 'square up', or 'by the wind', Mrs H gave him a certain look, and sighed:

'Oooh, just fancy.'

His Pa's greetings were a touch withdrawn and respectful towards him too, while his brothers' halloos were less easy than before: 'So who's this sailorman,' they mumbled, 'that's come to terrify his nephews with tales of fearsome sharks?'

Somewhat a stranger, it seemed, who brought a carved coconut showing the ornate N of Napoleon Boneyparts; and pairs of sandals woven of rope, that nobody knew how to attach to their feet, and when they did they fell over; along with whalebone trinkets shaped as African heads—a man's, a woman's, a child's—that when inverted turned into neatly carved private parts. Nothing like them had been seen in a Covington household before, nor in any Bedford Congregationalist's house either, except those of backsliders and harlots. It had not occurred to Covington that the roughness of his brothers seen from the angle of returning home might be changed, and appear confined and pious to him in the new eyesight he had. He was as puzzled as they were by such turns of events.

'I thought my gifts would make you chuckle and hoick,' he told his big-handed brothers, who rolled the objects in their palms, revealed them to Mrs Hewtson—who squealed—then closed their hands around them again when their good wives came into the room. You would think their youngest brother was a pagan, the way they pursed their lips at him. But when they took Covington to the inn, and all drank together from frothing jugs, nothing mattered so much any more. After Covington played his Polly Pochette they hoisted him to their shoulders, rode him out into the chill night air and dunked him in a horse trough, and he was a boy again for all time.

Mrs Hewtson reacted to his changes in her new fashion of uncertainty. While her affection was strong her playfulness was guarded. She kept telling people: 'He may become an officer one day, he's been associatin' quite grand,' which shamed Covington for his petty boasts and misleading hints of advancement. But after a few days her spirits improved and she became like her old self.

'I know what they say about sailors, Syms dear. That the whole lot of you are hard to keep down.'

'That we are,' he winked. 'But in what way, Mrs H?'

'There are places where sailors go,' she waved a pink, chapped hand vaguely in the air, 'you know the kind of house I mean.'

'Cunney-warrens.'

'What?' she blushed.

'Hummums,' he smiled, getting her going. '*Stews.*'

She pretended vagueness, and sighed. 'I surely don't mean anywhere like that.'

Covington widened his grin a little sarcastically. 'You do so too. But we had that good man aboard, John Phipps, who kept us clear of sin.'

'No man is ever as good as he claims.'

'Yea, you are right about that.'

He thought of Phipps's jealousy and hesitated before continuing. Had he done wrong? Yes, in the eyes of the godly, but nay when he kept in his mind what was best in his mind. So he said: 'I never strayed the way you think.'

'What, you mean you have kept your heart for someone at home?'

She pinched his cheeks like she used to, hinting it must be her, and he turned aside and found himself blushing like he never had before.

'Nay.'

'It is nay to everything, with you.'

'Well, I loved a maid, but it was in no pleasure-house. She was black.'

Mrs Hewtson's eyes widened.

They went to their chapel and sat with a small but

distinct space between them. Covington was a little ashamed to twist his neck around and look at the stained-glass window again. He remembered a day when he and Joey had stood in the great cathedral at Rio, still only partly finished. It had Marys taller and slimmer than poplar trees and brocaded in gold leaf worth many cargoes of goods. They made this chapel window seem paltry and perhaps a mistake, the outcome of mean resentments against the grandeur of Westminster and Rome. Covington wished he were Anglican at least, to have something like full beauty at his call.

He had come to a point of valuing the good workmanship of things, and recognised that the window was crudely made. The lumps of lead piping joining the glass were thick as sausages, and in the time Covington had been away had come apart and exposed holes to the outside air, where spits of rain blew in. The elders of the congregation who had ordered the window from London were mean penny-pinchers, he thought. The leaping young man's thatched golden hair as he cleared the stile was composed all of scratches, and his golden buttons were just splodges of paint. These imperfections he had never noticed before.

But at the call to prayer he clenched his eyes tight and strained the muscle-cords in his arms and shoulders, making himself strong in the House of the Lord.

After service he heard from their preacher that Phipps had been through Bedford. He'd come past with a small disciple in tow but now they were gone away. 'That will be my Joey with him,' thought Covington with a smile. 'They are headed for that chestnut tree. All will be well if we meet there.' He was stung, though, that he was not sought out at his father's house or butchery, and spat in the gutter a solid gob of envy, blaming Phipps, picturing him tugging Joey by the wrist, as he had seen him do, and dragging him over these very cobblestones away from Covington's presence, for the simple reason that he had opposed his will sometimes.

**N**ights were getting cold. Leaves had turned and were starting to fall. Gales shook the trees and spread the lanes with acorns. Covington had an instinct to meet John Phipps as an equal on solid land, and have them own to each other their true friendship. Covington believed something could be rescued from their old affection—and he knew that Phipps in his pride did too: both longing for the peace they had shared for a time. It ran deeper than hair-splitting evangelism. He wanted to say, 'Let us all three start our walking again, and find another ship.' Phipps it was who had propelled him into the leaky heart of the *Adventure*. Phipps it was who had given him his start at being a man.

'Then why do you niggle about him all the time?' asked Mrs Hewtson. 'I hear nothing else every time you speak his name.'

'I am not disloyal,' he told Mrs Hewtson, 'but I must be truthful to myself.'

He told her of the vale he wanted to visit, where he thought he might find John Phipps and patch their quarrel. He thought of it often—a moon hanging in the bare trees, the distant barking of a dog. 'We prayed under a chestnut tree, in the open air, like they did in olden times when our chapel was outlawed in the Restoration.'

Covington decided to set off there, but did not go yet

because Mrs Hewtson would not hear of it. 'We are having you a while,' she told him. She sat him around firesides as he told tales of going about in snowstorms, on seas of freckled ice. After the first shyness he was boasted elsewhere by his Pa, earning swigs of grog for telling his story in inns: how those wild Patagonians that they had sailed among were a people bewitched—scarcely like earthly beings and who slept in grass nests; how when they talked they frothed at the mouth, being excited beyond reason; and how when they wanted a destination they ran so fast their noses bled. 'I have their words on catgut,' Covington declared, 'having written them down for our surgeon,' and made the tick-tocks of Fuegian speech in pizzicato, *wen* meaning duckling, *i-ish* a feather, *appubin* a fish, *tomatola* a small fly to be killed with the smack of a hand.

A sailor could keep an audience of countryfolk spellbound for hours. Covington had not met with any sea-monsters yet, but people were just as strange. He told how their ship's surgeon had done measurements of natives, getting a few up onto the deck. Wide foreheads, smooth bellies, stout legs and broad flat feet they had. The men had little beard and no whiskers. 'Are they *hooman*?' a doubting listener wanted to know.

'Well, me and a sailor saw a maid of theirs up a cliff,' he answered. 'We went to see if she was made like a person, and aye, she was. Except my pal and self had no agreement on the outcome of the matter, for he snorts while I stare, "I would not go to any Fuegian for *my* fresh greens, shipmate."'

A Romany man heard Covington's Polly Pochette and proposed a fairground tent: the Musical Maltoot, he envisaged, Come Hear Him Play!—offering bed and board in his old wagon, where the young sailor might sleep on the same large bed as himself and his stinky old wife. Like cack he would! 'I will not be stolen like any crazed Patagonian eating sea eggs on a plate of ice,' he told his brothers, 'dressing for dinner in orange paint with kelp in my ears.'

As for himself, in the moan of his fiddle, Covington remembered how squalls came through the narrows, ruffling blue water and dipping tree branches in waves. The pitchy darkness and the whistling wind were in the catgut. As he curled for sleep a memory shot through him of glistening wet eyes, a wide forehead, smooth belly, and plump breasts like pouches of sand. He thought of her as he bunched himself down for sleep. Who on earth was she to bedevil him so? He followed her through water that was not cold, in his memory, but heavy and sweet as honey, in which she was held, and willing. At the last moment, as he spumed his load, she looked at him and did not turn away.

Mrs Hewtson lifted a tattered blanket whereunder lay her newest sprogs, dubbed Spit and Polish. There was less room than ever for Covington's knobbly knee-bones on the sleeping shelf, and he fell to the curdled boards with a knock, ready to be gone.

In that deep October there was rain, but then it cleared and the whole county was misty and still. Smoke hung over the town in a close, thick band. Above, the moon was a brilliant splinter. Covington went walking the countryside with bread, cheese and mulberry wine carried in his satchel. He hefted a blanket roll, too. He plucked ripe pears from overhanging branches. Away from towns the stars were brilliant, sharp, and he imagined running his fingertips across them, feeling them rough as grains in holystone.

There was a sailor's swing to his gait and it seemed he could hear the roar of the sea at the edge of his hearing, tempting him back. It was like a reef moaning under the drop of the horizon, and it played in the key of G. Falling asleep, he missed the pad of the night watch on the deck, the complete organisation of everything. He pulled his blanket around his shoulders and wrapped himself in it and burrowed into haystacks. Morning, he was glad to be alive.

There was money to be had from digging ditches and clearing drains ready for the winter floods. A few shillings found its way into his pocket that way. He was a great worker, an easy companion. But there was always somewhere else he needed to go.

One day Covington crossed six low hills, counting each one, and came around the corner of a ruined barn and found himself overlooking the shallow vale he longed for, and the remembered chestnut tree. It had been bare the last time but now was in its last full leaf and turning gold. The spikes hung like tassels on a curtain.

He stood there in wonder. The tips of his fingers tingled, and he wanted to reach out. But what to touch? The tree was a throne in his eyes. The air was luminous around it, golden with autumn light. All creation sang in the presence of a maker who seemed, when Covington turned around quickly, to be invisibly standing at his back and smiling. The roots of the tree made places to sit, benches and settles and a soft couch. The leaves made an arbour above Covington's head. A man starved of God had stood there, casting around with his fierce eyes and making a plea. Without Phipps to impose catechisms Covington made free with his own. By the going up of the tree his thoughts went up to heaven. By the light of the sun beaming down he thought of God's understanding reaching him. By the roots of the tree, finding a spring, he considered his own deeper nourishment. He thought how the spirit was never seen, but felt, and so was known. Thus he ministered to his own needs roughly, clumsily, without instruction, as he was born to do.

A voice inside him had always said, 'Make something of yourself.' Now that he had a trail of life behind him, Covington saw how the path wound around and came back upon itself.

The sun went down and he saw Joey Middleton and John Phipps coming from the distance, striding inside an inky shadow. But there was nobody there, though he ran to greet them, deceived by old gnarled roots and bare stones.

Later he lit a fire, wrapping his blanket around his shoulders and listening until his ears ached to the distant sound of shooting inside the walls of the park. The gentry were going all at it for snipe in the woods over there and Covington could well picture King and Musters trooping along with King's father and his brother-officers in such a place, for it had been their plan to go shooting when they left the ship. One half of their life was navigation and the other half was creature-carnage.

But then watching them in imagination Covington also pictured John Phipps with his thin, knowing smile setting snares and acquiring his part of the bag for the sake of the good idea he had, that the countryside must be free as the ocean, and that a man, choosing God, was unfettered before the wind.

Back when they had signed over to the *Adventure* Phipps had taken it for an omen, because of the commander being named King. 'Those named King know humility in their bones, for they are fallen from high places as their family name indicates.'

'What has this to do with being a Christian?' the boy had argued.

'Everything indeed, because those who are beggars may also become *kings* in heaven.'

Covington smiled at the memory of all the tangles Phipps made in his thinking just to squeeze a parable from a situation, or to make an interpretation convenient to the day. It gave comfort to an ordinary soul to hear a righteous man bedevil himself. Covington knew that Phipps had just liked the look of the *Adventure*, and felt familiar with her lying there in Monte Video waters because two old shipmates, Door and MacCurdy, had already rowed across and were snug in her, and Covington did not think it such a bad hypocrisy to have, either, in a man who loved the sea life.

After it was fully dark, and misty with starlight, Covington went down to the chestnut tree and prayed. In his aloneness he felt as if he rolled a stone from a door inside himself. A feeling shot up from inside him akin to light. It was such a powerful longing that it made a shape in his inner eye as he pressed his eyeballs with his knuckles. And who was standing there, robed in white, gesturing him to follow? It was the son, Jesus of Nazareth, who had trodden the dust, drunk the water, touched the leaves of the tree, and had been in every way a man before he took his step upwards into glory.

Covington scrambled back to his sleeping place in the ruined barn. What had he seen, if anything at all? He lay with his chin cupped in his hands, staring out into the silent night. He saw the white-robed man running up a rocky path and dashing along the stone wall and into the park. He smiled. No—it was because he pressed his eyes tight while praying. But Lord did he have such gratitude for just being alive.

He sang himself to sleep with Bunyan's hymn, that was their anthem in chapel and always brought tears to worshippers' eyes:

*Hobgoblin nor foul fiend*
*Can daunt his spirit.*
*He knows he at the end*
*Shall life inherit.*
*No lion can him fright.*
*He'll with a giant fight.*
*He'll fear not what men say,*

*He'll labour night and day*
*To be a pilgrim.*

The next day Covington returned to Bedford where, at his last farewell, he gave Mrs Hewtson a great kiss, took out his Polly Pochette, and stood on a chair and played 'Greensleeves' in his sweetest style, while tears ran down.

Mrs H said, 'How shall the young sprats live, if you don't bring a fortune home?' She cut fresh plumcake and wrapped it in muslin and sealed it inside a tin. 'Take this with you, mind.'

Long-stepping, sizeable as a man, Covington walked to Devonport in bare feet, economising on shoe leather. When he was almost there he saw two figures crossing a bare field ahead of him. He fell in behind them and heard they were catechising each other. It made him smile, grin and clutch his heart with joy. What fantastic simpletons they were, the doleful sailor in the battered, three-cornered hat, and the sprightly, bare-headed sailor boy at his side. How ridiculously easy to dog their steps and hear the way they did it:

' "What is the fear of God?" '

' "The fear of God is the beginning of wisdom." '

' "What do they have that want the beginning?" '

' "They that want the beginning, which is fear, they have, they want," ' stumbled Joey. 'Blazes, I am too tired to know what they want, Phipps.'

'We are almost there. I see the town,' said Phipps, pushing the boy along.

' "They want ... " '

At this Covington leapt in front of them, and spread his arms wide, stopping their progress. ' "Those that want the beginning have no middle or end!" ' he bellowed.

'That is it!' yelled Joey, throwing himself at Covington and punching him on the chest. 'Where have you bin gone, Cobby?'

Covington held Joey by the thin shoulders and spoke across his head.

'What have you been feeding him, John Phipps, weeds and wildflowers?'

'All that I give him he devours.'

'Only I choke on them catechisms,' said Joey, kicking a stone.

Phipps embraced Covington and called him brother.

'I am glad to find you,' said Covington.

'And I you,' said Phipps.

'I was always coming to find you.'

'I like the way you came along.'

'I would have bit a firebrand, had it stood in my way.'

'I am glad to hear of it. We waited for you under the old tree.'

'Two nights,' piped Joey.

'I was there, but not soon enough,' said Covington, feeling ashamed of all his resistance.

'Soon enough is the time you were there,' said Phipps. 'It is a great mercy-seat under that tree, is it not?'

'I'll tell you what, I prayed,' said Covington, 'and there was no way around it, John. I was given to understanding.'

'Can you say what that understanding was?'

'Faith.'

'How do you explain that faith?'

'There is no explaining it.'

They passed through a farmyard and drank from a watering trough.

'Well then, I have heard of a ship,' said Phipps, clapping his spindly hands together. 'She has a Christian captain and I am sorry to say no maid for a figurehead for you to moan over, but only a dog.'

'What sort of dog?'

'A beagle-hound. She's our sister ship, she's had new mahogany fittings made and has been all done-over for her next voyage. We are old Patagonia hands, believe me, and they will take us if we want. They are all for putting missionaries on Cape Horn.'

'Where have you heard this from?'

'From Door and MacCurdy.'

'What? Have they repented?'

'Nay, but I have my hopes,' said Phipps with a grin, his sharp Adam's apple wiggling up and down, his parched lips and feverish eyes fixed on Covington in liking.

Within a day they sighted her. Their seafaring language, coarse and hard as old rope, returned to their speech and made them ready for rowdy company. Their catechisms crept away inside them for the moment, like snakes that were shy of the cold. She was tied up alongside the wharf with the look of a captured cockle. Carpenters worked on saw-holes in the deck, creating storage space after raising the upper deck to allow more head-room. Wood shavings floated in the water and sawdust blew in the wind. Cables were strained, and crates of rations—beef, pork, peas, vinegar, rum and cocoa—were carried below by gangs of sailors shuffling one after the other. They remembered her from the days of Captain Stokes as a rotting miserable tub, a former collier in the coastal trade, except she had faced the roaring nowheres brave enough (braver than her captain, who had stiffed himself). Now she was being given three masts and smartened over, being changed from a brig to a bark, though nothing could make her bigger—she was fearsome small—just two small cabins, with the dead commander's space rebuilt. She was ninety feet long with a beam of twenty-four feet. They could not say they loved her, for they had no warmth of life yet shared in her. Yet they loved her promise.

They stood, the three of them, at the dockyard gates, feeling awkward and looking quaint in their patched and

torn clothing and straw in their hair from sleeping in a
stable. The watchman greeted them in disbelief. 'Crew for
the *Beagle*? Watch out your bosun don't see you. You're
more like mushing mudlarks with shit up to your armpits.'
They went to lodgings and took a bath. The next day they
returned through the town to find the vessel shifted from
the wharf and riding free in the water, and smarter than
yesterday by a thousand degrees.

A longboat took them over the choppy grey water and
they scrambled up the side. Their seabags and prized possessions were hoisted by a whip from the yardarm. They
eyed their new companions, a set of blue-jawed, cavernous-
faced piraticals. One had a bamboo flute, another a small
skin drum, while Covington had his fiddle and so there
would be music made. Joey Middleton was first onto the
deck and gave a whoop, sighting Midshipman King, calling
him 'Phil' and striking him on the chest familiarly.

King reacted with only a faint smile and Covington knew
better now, wincing as Joey was told: 'How are you, boy?
I am afraid you must call me mister. And please, your pea-
jacket has a smell about it.'

'But I brushed it well,' chirped Joey, and it was a shame
to hear him carry on. 'Ask Phippsy.'

The midshipmen were all bunched in a pack, draped on
the foredeck appraising the new arrivals with their caps
pushed back. They had their clay pipes between their teeth
and went puffing away.

'Howdy-do, Covington. Is your mother well?'

'Very well, King, and yours I trust the same?'

'Middlin'.'

'There shall be no joy here,' thought Covington, as King
continued:

'Middlin' for a woman who hates the sea and now must
join my father in Australia.'

In the scramble to be of best service Joey won the day.
He was made servant to the poop cabin where they were to

welcome a wealthy young gent aboard, a Mr Darwin of Derbyshire, twenty-three years old, a bug-catcher and very close with the captain in all his dealings. To be on the good side of a well-heeled passenger, it was agreed, was a very fine thing. He was counted upon to have silver to spread around for favours.

'You are quicker than hell would scorch a feather,' said Covington to his small friend when he heard that Joey had won the gent.

Joey dropped his eyes—apologising for favours he hadn't sought, but were granted him for his smooth cheeks and merry eyes. It wasn't his fault that a nipper was still a nipper and so made a pet as affectionate as a marmoset. Anyway, his gent wasn't aboard yet. He was sleeping with the captain in lodgings in Devonport. It was the captain who had put him forward.

Covington, with his strength and willingness, was made ship's fiddler and odd job boy. It meant constant mending of broken wood, also tending to chickens and goats, and jumping to command whenever a need was quick, bringing mallet, twine, and tar-bucket at any hour; and if there was need for a dance, to warm his fiddle with a jig. It was no change from before but was only his new start. He longed for better and plotted to put himself in the way of the captain, to display his penmanship and willingness, and be preferred as Joey was.

They hung about at anchor, awaiting favourable winds. The season turned blisteringly cold. They got the measure of their Captain FitzRoy. He was a wind-chapped, peaky-faced young aristocrat who would have everything done before he thought of it, with a habit of placing a finger to his chin and staring with glittering eyes to etch his memory. Nimble Joey, with eyes of a fawn, polished silver and folded napkins without being asked. At the last minute as the wind swung to the needed quarter Mr Darwin came aboard, and Covington, at the other end of the ship, gained no strong impression of him that first day, except to note that he was tall, round-headed, and as dully dressed as a curate. The way he jumped back whenever a sailor ran past denoted a landsman's willingness to apologise for his existence, while at the same time ensuring that all his baggage be placed where it was most likely to trip a sailor over.

Covington stared at the missionary they were carrying to Cape Horn. He was proof that the Lord does not choose those that love him. Revd Matthews was a smiling, sly man of God, mawkish and sweet of manner. The three native Patagonians who came with him were another story. They were smoothly at ease, dressed in frocks and topcoats. They behaved like proper Methodists in allowing themselves to be meekly marshalled below for supper and prayers, but their eyes flashed around with mischief in them.

At the cry of 'Man the windlass!' and the sound of the bosun's pipe, FitzRoy in high spirits shouted loud as if all voyaging was new to him: 'Ho for the Canary Isles!'— which was to be their first anchor on the way to putting their human cargo into the wilderness. All the passengers were sick, and not to be seen. But forthwith the *Beagle* ran aground on a mudbank and they were called onto the deck to add their weight to the port-side. The wind howled as all seventy-three souls paced from one side to the other, and rocked her free. Humiliated they returned to harbour. The crew said darkly there was a foul-weather Jack aboard to cause such trouble, there had to be someone and maybe it was the gent—that flap of brown cape and pair of white hands gripping the side rail as he spewed and spat before disappearing inside again, sick as rotten vinegar. Joey threw a punch that bounced off Covington's chest: '*Covington* is the foul-weather Jack,' at which Covington growled and fended Joey with his palm.

In regard to their gent there was another thing Covington noted, a quality in him remaining true years later: that while he was faint against his background he was strong in his effect. Phipps worried that the gent's father, Doctor Darwin, being the richest man in Derbyshire, was surely a great patron for a favoured boy, and he ruffled Joey's hair as he said it, fearing to lose Joey, his chosen one. Covington, with narrowed eyes, considered his displacement in Phipps's affections, and allowed for Christian forbearance to get him by.

More delay, and then full readiness again. They were tired of the sight of the town looking miserable in all its shoregoing dullness, with damp slate roofs, blackened chimney pots and driving hoops of rain. Then wind caught their bark, and she was away again. Out into storms she ran, bucking like a burro. With much hauling-in of canvas and landsmen spewing and no distance made day after day, Door said, with a great sneer: 'There is someone ashore keeping a black cat under a tub.' So turnabout home to Old England they steered, unable to fight the moods of the planet. They did not go, and would it have been better if they had never gone at all?

With the ship at her moorings again and looking likely to stay, the young gent went ashore for shooting in a lord's park. Covington saw his slightly stooped, broad back, his gun case and powder horn, and felt a pair of rounded eyes pass over him, seeing nothing at all. Or seeing, at best, the world as it was arranged. You could love the naval service and be part of it, and still be offended to see how it made a landsman smug. *Mister* King was heard to say he would have a good time of it now, and while Covington went about with a bucket and rag wiping spew from mahogany fittings in the poop cabin, King slipped ashore with a gun of his own.

Covington kept a hangdog attitude as he swabbed. Christian forbearance was all very well but the poop cabin

was not his part of the bark, he was lodged in the forecastle above the coalhole. It was Joey's place there, aft among
the toffs, Joey who waited on table and was offered
marrow bones to suck and plum pudding in hefty slices,
Joey who softened Capt's Christian eye and made the
Patagonians they had on board giggle and like his sweetness and true affection.

Joey showed Covington nets, guns, microscopes, tele-
scopes, hammers and tongs. The first solid bit of their gent
that Covington ever knew was yellow bile on a letter case.
Joey said the gent had a plate of Wedgwood's pottery
showing a black man in chains, and the words, 'Am I Not
A Man And A Brother?'—at which responded Joey with
tears in his eyes about the fate of slaves, '*El pobre se siente
intimidado*,' for when he wanted to speak from the depths
of his heart and not be laughed over he chose the Spanish
they learned on La Plata with Capt King. There was
another reason, too, thought Covington.

'You have been lying in brother Phipps's arms, and he
has been a-whispering in your ear, and lo he has made a
slave of you to his affections.'


There was nothing for it, for a time, but chores and duty as
the *Beagle* was going nowhere. Indeed the whole Royal
Navy went nowhere, staying clustered in harbour like
clotted leaves on a pond.

A portside hatch needed painting. They worked amid
shouting, curses and laughter, with bursts of rain on the
tar-slicked timber. A gadabout gull came skewing down
the adverse winds to defile what it could, extending its
claws, wings flapping, and made a great squirt. '*¡Hola!
¿Como estas?*' They would have no more of its foulness as
they boasted their lingo. Covington shoo'd the thing away,
his fingers trailing on feathers as he lunged.

'Had 'em, Haddums!' cried Joey to its screeching face. 'Does your mother know you're out?'

That was when their gent grunted up the side and for the first time in all creation met Covington's eye—the boy registering a round coppery face and lubberly sea legs—one, two, and a clumsy haul, and Covington had his man to observe, all the height of him uncoiling shy. All he knew of him at present was that he liked to go out with his gun and his dog in the rain. He was, some said, a young squire of the sort who passed time with philosophers discoursing on whether Greeks ate melon seeds, or if they had privies in their gardens. He came from dockside in a cutter near sinking under the weight of extra goods that he wanted this late, everything awkward-shaped and dripping in the December mist as it was hoisted: a bundle of guns, a crate of jars, a sack of books, a rectangular basket lined with paper that was meant for dead birds. As he wondered, 'Might he trouble them with his extras?' Covington held his gaze and heard the words, but the gent's brown eyes still looked through him. 'I am ashamed,' thought Covington, 'to be who I am.' His way to counter that shame was a sudden whim: he would get the gent's attention and if need be wrestle his service away from Joey.

It was at this moment there came the sound of a splash that Covington would never want to hear again, as when a bucket of swill is tipped overboard in a calm, and a small voice in the heart of him cried, 'Joey? Was this Joe?'

Yes it was Joey slipping from his rope and falling in the water.

Yet even while thinking this thought, so shrill with alarm, something kept Covington staring at their gent, trying for a response.

Joey Middleton dear sailor chum unable to swim, where was he then? At Covington's feet with his fingers clasping holystone, he wished. But really where was he at all? Swinging above the deeps lashing at gulls with a coil of hemp? Then what? Blood on the shit of the wind! *Drowned.* Covington's attention was elsewhere for the count of three. For he was *still looking at the gent*, musing on being of service with a warmth inside him akin to desire.

'*Boy* overboard?' FitzRoy burst from his cabin under the wheel to roar like a furnace. Yes with your sharp nose like a blade and your fishy lips you may have your spectacle, Capt, of a soul entering heaven.

MacCurdy dived, Door went after him, trod water and hoy'd. Phipps's arms pinned Covington to the rails to stop him leaping. 'The two of them are enough. More will get in their way.' MacCurdy broke the surface, dived again, returned shaking his head, 'He is *gahn*!' Covington trembling aghast, Gent likewise mumbling a prayer, plucking at shirt buttons in the icy wind, the crew gathering, climbing every vantage point on the ship, bewildered, shivering, delivering the news to each other in undertones. Phipps, meantime, what was he thinking, his eyelids tremoring, his mouth tight shut, surely also of heaven?

'Gahn!' came the bellow again.

Capt echoed the sour word in disbelief and grim-featured called for hooks. He could ask for no better crew than a hydrography ship's to plumb the depths of a harbour. Oarsmen dipped by and held. Dipped and returned. Men able to tune the wind and furrow the four seas grieved for skills denied them in the matter of a small life.

Joey was down there a long time.

'He is gone to his father,' said Phipps, grabbing hold of Covington again. 'I knew it would come. He was too sweet for our world.'

'Let me go, John Phipps, or I'll use my strength on you.'

Phipps was sobbing, poor man, and there was nothing for it except to turn aside from him lest Covington's own sobs crack his bones.

'We have summat!' cried the West Country men in the evening light. Covington clutched a rope and began hauling. They brought Joey up. His lips were nipped by fish, his sides were torn by grappling hooks, his clothes were peeled away, his eyes were hollow sockets!—he was upside down, dripping like a sponge, and their angry FitzRoy never spoke a truer word as when he boomed out in the presence of crew and weeping Patagonians, 'Lord God of Hosts, Joseph Middleton was your *friend*!'

'Amen!' answered Phipps. 'He formerly lived by hearsay, now he shall live by sight.' But all the texts in the world were clay in the mouth and thoughts of heaven collapsed, and Phipps wept, unable to get the words out.

Covington fled the *Beagle* and ferried ashore in the dark.

''Ullo, it's the mudlark.'

The nightwatchman with his lantern on high signalled him through the dockyard gates.

He carried his Polly Pochette with him as he ran, holding her by the neck, her dry dusty bow thrust deep in his inside belt. Sailors heaped bonfires in the wet streets and swigged rum. Covington wanted no more of their game. A wild gang passed by, grabbing men. Covington huddled against a chandler's warehouse, listening to his heartbeat through the walnut space that was no bigger than his heart itself and as flimsy. Then he ran on. The rain stopped. There was nowhere for him to go. He sat on a stone and plucked strings, and wept. It began to rain again. In spite of the horror of his dear friend being flayed by water he would be shown the cat on his return. He knew it and returned in the miserable pearly light of morning. Their Capt eschewed exceptions. He spake the law of God. Covington could rely on him.

It was said that in fine weather, sails bellying full, there was no joy a seaman knew better than departure. He was rid of the woes of the land, the ship heeling nicely to leeward—the sea washing her decks in a torrent of green—the steersman heading her up to the wind and ever into warmer climes.

Covington considered this without bitterness as he bared his back. A brightness sparkled on the water. He was given a rag to bite upon. 'Good cheer, Cobby,' said a mate, and Covington gave a smile in return and shouted before his mouth was stuffed:

'It will be all the same in fifty years!'

As he submitted to Regulations he saw the gent watching and taking pity, hand to mouth to hold back spew as their bark went on. Covington did not allow himself to have any thoughts, then, except just one, that he was no dumb beast and the pain he would fling from him like water, because it was his temporal body, that's all. The gent was braving himself in FitzRoy's eyes as one who would not speak out loud, but hold, as a guest aboard must bravely, to naval lore in a flogging. Phipps stood by with the others—his own back was scarred, written plain with the transgressions of his youth. Covington till now had only ever suffered tongue-lashings, and boxings around the ears. Other thoughts crowded in. He thought about the sweetness of

Joey, blaming himself for his drowning—those few moments of neglect, of not seeing what was happening. He might have pulled him back. He might have said, 'Don't, Mudlark, Noodle-head,' and gone for his own turn at swinging above the water. He might have drowned in his place, and not had this bitterness on his palate, which was made of the taste of seawater, weed, stale cotton and shame. He might have averted his eyes from their gent, instead of coveting his service.

*Bring it down.*

The lash whistled and Covington was flayed for his shoregoing. A stroke for each hour taken without leave, making fourteen in total.

Captain FitzRoy stood watching, a severe young man with a long slim nose like a paperknife. His firm full lips were thoughtful even in a rage. Two times seven he eyed the rope descending and surely he thought of drownings, and knew of Covington's grief from his eyes, and his pain, but thought of obedience making all things right on ships, and surely in heaven, and so, to be sure, was soothed in his hard decisions around men and boys.

After Covington was done, his back sliced and his groin weeping bloody piss, there were the greater shore offenders to be dealt with, Christmas punishments for drunkenness making a stern tally of one hundred and thirty-four lashes in all, their cries piteous, their flesh lying on the decks in scallops as if from a carpenter's plane.

Covington was hurt. He stood aching by the ship's rail, swollen with his body's water. A bucket of saltwater stung him. The surgeon's mate applied linseed mash. He saw his skin hanging when he twisted around, donning his shirt. The surgeon, Mr McCormick, said he was fine. King failed to meet his eye. Musters bawled and hid his face. Phipps came to him and gave him a swig of rum. The other topmen left Covington alone; gormless whales of conscience, they regarded whippings akin to weather, to be ridden through and rewarded with fine passage after.

Covington nursed his body through the next weeks and ate Mrs Hewtson's plumcake in sorrowful nibbles. There was little conversation in him and no light banter. He moved in the vessel with a sullen power, answering no arguments, entering no pleas. From this date he stood, as on some fearsome headland, remote and proud, surveying his years ahead. It broke hearts to see him, for he was still just a boy. FitzRoy gave his Sunday sermon on the open deck, making his verse the rainbow of hope that Noah saw at the end of the Great Flood. There would be a gathering-in on this voyage and a putting out, he said. A spark would be lit in Tierra del Fuego and it would be a beacon to Christians the whole world over. The men hardly knew what he meant, or cared; they just agreed that if there were missionaries in wild places then maybe their chances of being rescued from shipwreck would be better.

Covington saw their gent fashion a bonnet from canvas and attach ropes to its sides, trawling it through combers, bringing up sea creatures, marvelling at their smallness on the deck. 'It creates a feeling of wonder,' said their gent to Revd Matthews, who was ever lolling nearby, 'that so much beauty is created for such little purpose.'

'Aye, but we see it now,' said Revd, 'and offer our hosannas.' The young gent turned and stared at Covington as if he were cousin to those jellies, and truly, as if to flatter his opinion, Covington was silent as a fish.

On Sundays, for an hour, he sat with Phipps side by side and strove for understanding. They riffled the pages of their book with Phipps tactful and withdrawn. He knew that Covington's entire soul and not just his skin was flayed—why, Phipps was no different—and they did not know or care if it was Joey or Christ they whispered about. The texts reminded them both ways:

*Wherever I have seen the print of his shoe in the earth, there I have coveted to set my foot too.*

*His word I did use to gather for my food, and for antidotes against my faintings.*

*His voice to me has been most sweet, and his countenance I have more desired than they that have most desired the light of the sun.*

Covington would not have been able to say, if asked, what he remembered about Joey to rouse such an ache in the heart. Memories were so slim, threadlike, that they seemed hardly to exist at all in the face of a life's passing. Joey was a boy who used nicknames, was honest about his failings, who danced lightly on the water, screeching 'Had 'em, Haddums!' before he fell. And that was just about all.

Mostly Covington and Phipps went to their different parts of the vessel and were strangers to each other. There was just this love they had, that was without its joyfulness any more, and was only sustainable in their private devotions when they asked God to take it to him.

**W**ord came of advancement. Covington was elevated to gunroom service, which was only halfway bad. It put him on the ladder to higher things and yet made him servant to those who once might have been his equals. He found himself running errands for midshipmen and the lesser officers whose social standing in the ship was a touch uncertain. His masters included King, the unreliable pal of his childhood, and so remotely detached from him now that he hardly looked up from squeezing his boils when Covington brought him a poultice one day. Covington had gone to great trouble to have it heated on the cook's precious coals, but King only said, 'Lay it on my knee, and stand back. Thar she blows!' and dirtied the clean deck with his blood-riddled muck.

Off Brazil the lads were made hilarious by a school of octopusses, as they called them, though Covington, who had become prickly and challenging in his moods, knew they were squids, flopping through the water making a poor sight with their tentacles chewed and their bodies in a palsy. It was done by dolphins, who after their attack stood off as if they feared entanglement in the debris that gathered around the ship's hull. Men mulled about in the strange humour of what they saw, for entertainment was all their life on the wave at that time, as the rudest mockingly called, '*Them's* not octopussies, but whales' dicks bitten by the moll of the deeps.' Mr Earle, the ship's draughtsman, smiled: he was a rare bird like all his artist-sort, and would play when boys played as if to be a child again, though he was the oldest man aboard. Covington had just begun to notice him—an easy, sickly, good-natured fellow with the knack of getting along with sailors and being as rude as they were if he chose. He thrust a pencil in Covington's hand one day and showed him how to draw two eyes, a nose and a mouth, then turned it over and the pretty young face was converted into a tree stump.

As for their gent, who had Joey lost to him as a servant barely before he knew who he was, he played at not knowing who he would have to replace him until the captain gave the word, and was blank to Covington's flirtations for the post. Their gent was a hard one to catch. He wasn't such a slug as he looked, even when he puked. He would dampen the wind, as they said, wipe his mouth with the kerchief he kept handy, and get on with his tasks forthwith. Could be he didn't need a servant at all. Sharp was the word and quick the motion with him, and he would not die laughing, as Mr Earle did almost, savouring every foulness he overheard. This Mr Darwin was something of a Scot in his dolour when the crew got going. There was just a deeper colour he turned at low banter—plum red as he sucked in his cheeks and went about his business. Yet his attention was constant. It made the hair stand up on the back of a young dog's neck.

It was good that Covington saw they were squids, for FitzRoy soon afterwards sent for him while he presided at a gunroom supper. Space was squeezed and manners cordial among these higher gentry. This captain was a great one for correction and often barbed his listeners with their ignorance.

'What were those humps in the water?' he smiled at Covington, holding a bearing on Mr Darwin who slurped his soup hungrily—for there was no motion that week to

make a landsman peaky. 'Those poor objects the men were
so engaged upon. I heard them callin' them *octopusses?*'

Covington told him straight, 'They were squids, sir,' as
he backed out, collecting a dish. When he was gone, 'What
is its name?' Covington heard asked from around the
hatch—and there returned the murmur, 'Syms Covington,'
confirmed among the company with yawns.

'Odd job man.'

'Fills in where needed and scavenges for what you will.'

'Brains?'

'An excellent copyist and a butcher's boy.'


One day Covington saw the captain moving along the rail
towards him. From his manner, thought Covington, he
might be one of those dolphins when it raised its snout,
finning along steady with a round eye on the lookout. The
gent followed him. FitzRoy went absently fingering the pig-
tailed scalps of sailors and gave no expression or indication
of mood until he reached Covington's place.

'A moment of your attention,' he spoke to the gent over
his shoulder. Covington felt the pads of FitzRoy's fingers
move across his skull seeking lumps and bumps. 'Now this
one for example. Covington, will you hold steady, lad?'

It was not a question that needed answering. FitzRoy's
fingers were quick and spiky, jabbing and plucking at hairs.

'Amativeness—full. Philoprogenitiveness—full. Concentrativeness—ditto. Adhesiveness—full. Combativeness—large.
Destructiveness—very large. Constructiveness—small. Secretiveness—large.'

'Am I a dumb ox,' thought Covington, still with his head
down, 'to be used like this?'

Then the gent had a turn at him and it was not such an
unpleasant feeling—soothing and distancing, rather, as if
one's own self had no case to argue. These hands were
already well-used, such mighty golls for a young gent of

twenty-three, rough-grained and white with starching or whatever chemicals he contrived from his pickling jars and packets of arsenical powder.

Then the hands were gone and the attention of the pair was elsewhere. FitzRoy murmured thanks but Gent said nowt.

Covington knew this much already about the one who treated him deaf as a mainmast. Naked life was his subject in nature. Rocky islands and stones from beaches were living materials to his eyes, likewise skeletons of fishes and birds from which he conjectured former existence. With his geology hammer it was knock and it shall be opened unto you. Whereas their captain saw weather in the skies, their gent saw weather in the ground, and in the rocks of the islands, and talked of changes to the earth as if what was placed by God in enduring stability was a theatre of sorts. The thought opened the eyes of anyone who cared to consider it. Covington was one: it increased his amazement at creation. Phipps was another, and smiled in wise agreement. God's book was a fatter volume than he had conceived. All earth spoke God's praises.

When Covington heard that his adventure on St Helena island was known to the gentry, and he was said to be advanced with the opposite sex, he noticed the gent's eyes on him more often—though without much interest in engaging him in conversation. Indeed there was a stronger shyness about him then, a withdrawal, if that could be said of a gent who barely came forward in the first place, whose attention clamoured from a distance.

They found themselves becalmed. It was hot weather, thick as steaming flannel. Passengers and crew mingled, seeking every patch of shade. Men jumped overboard and swam around the vessel to wash themselves. Covington played the fool and acted sick for Miss Devil's Horns every time he saw her. He had a bad cough just then and was somewhat feverish. Horns was the female of the Patagonians, Fuegia Basket being her Christian name, fourteen years old and sitting on the deck hoping for a breeze. Covington made out he could not keep his young dog's eyes away from her. The lads said he was lost in Fumbler's Hall, and clearing his throat he said, 'Sure and I am struck in the philoprogenitives real hard,' hitching the crotch of his trousers.

Then he had a dream that made him look at her twice. In the dream she reached out a hand and drew him to her. It made it seem she was the one who would rise from the water some day, riding in a seashell in all her Sunday School finery and savage guile. It made him go around looking at her closer when he got the chance. One day when he stood near the dame, and roughly catarrhed his lungs, getting her attention, Revd Matthews came by and spoke of Miss Basket having a *novio*, as if underlining his words.

Covington raised one eyebrow and said, 'Who is that

intended you mean?' and Revd Matthews nodded in the direction of Yorker, another of her tribe, who played with his knuckles as if he would pull bones from them, and was glowering at Covington something fierce.

'They shall be joined under holy law,' said Revd, who called Capt 'his Nazarene' and made listeners retch. It would make Covington fling a punch if ever anyone took *him* for a Methodist.

They longed for a breeze, a diversion. The dolphins returned. The lads named the leader that still followed the bark Malty, from his drunken cavortings; another was called Do-Little Sword, after a midshipman's dirk; there was Pin-Buttock and Quatch-Buttock, on account of their thin arses; and the smallest that lingered behind was called Come Out of That.

'He is a right shallock, a dirty, lazy fellow,' said Covington. 'Malty is the one I like best.'

'I'll put my shirt on Come Out of That,' said Mr Earle, with his sketchbook held open on his lap. 'He suits my style.'

'*Your* fish,' said Covington, 'is lazy as the tinker who laid down his budget to fart.'

MacCurdy and Door started putting tobacco and rum on who would gallop around the ship first—it was always Malty, slick as a fist in a muslin wave.

Earle did a likeness of MacCurdy, a fearsome seaman bared to the waist and wearing sailcloth trousers with a careless sashband. He was first done in outline and then Earle rattled his brushes and gave him colour, and it was like a rainbow filled MacCurdy in, while the waves and decking around him stayed dim.

Something in Covington's thoughts turned his head around. The gent was at it again, assessing a person while seeming vague. He looked at Covington, who was without his shirt also. Gent dropped his gaze forthwith, getting an eyeful of pustules that were still in their last stages of healing (and giving the boy lumps under the arms from the poison in them). Covington would not let his gent go so easily, and with wheezy lung baited his eye.

'Would you have somethin' for a cough?'

The gent nodded, then turned on his heel and made for the lower deck. He returned with a phial of eucalyptus oil. 'Try this on your handkerchief.'

Covington produced a rag, 'My thanks,' and trickled the stuff on, and inhaled it deep.

When he was quite well again, with tropic air and salt water doing their part, he went to the ship's library in the poop cabin where officers were told to go before eight-thirty in the morning, if they wanted a loan of books. There were hundreds of volumes close-packed, among them the *Encyclopaedia Britannica* in twenty volumes which Mr Stebbings pulled down for him with good humour when Covington said there was a wager going on over who was the greatest fool among them.

'Bumpology, is that what you want?' asked Stebbings, and Covington said, 'Yoi, whatever'll do'—and he conned the entry quick smart. It was like the brain had several parts, each representing a faculty of thought, such as love, morality, veneration, greed, preferment, lasciviousness. He knew that his own head was a lumpy job. He knew which qualities gave him an itch in the belly, and which made him whistle with hope and pleasure when he woke in the morning and jumped to his duties after saying his prayers. Between the bumps on his head and the blood in his veins there was this third capacity making him glad. It was the spirit of the Lord, the joy of being alive. He would be robbed of his meaning without it.

The night was close. Along the deck sailors sprawled in all manner of comfort, some snoring, some singing. The officers were drunk. The bark swayed as if at anchor on the smooth plain of the sea. When a breeze wafted across the water the officers scrambled on deck and cooled themselves, flapping their shirt-tails like chickens, and then they marched below again, spouting nonsense.

Covington waited by a hatchway for orders, and he slept. Then his name was called and he was in the galley and into the gunroom smartly.

They were getting up a good chatter in there. Four of them were close-packed dining and gossiping. Covington took up a post within earshot. Soon enough he came round as a topic of conversation, for he dropped a pan, spilling haricot beans, and made a bother of himself. Mr Earle knew Covington as one who liked to draw. Mr Wickham, the first lieutenant, traded on the merit of Covington's eye for 'that fiendish little bit of frock'—he was, said Wickham, 'admiral of petticoats'. All laughed. FitzRoy hurrumphed. Darwin said nowt. FitzRoy then said words that sparkled in Covington's ears: he said he knew Covington as one who might serve any man as a clerk—that Capt King had taught his boys Penmanship and Style as he surveyed La Plata waters amid eddies of weed and mud, and King would often test them with spelling and grammar, and this *admiral* came up trumps. 'The *Adventure*, why, she was a very Dame School,' FitzRoy brayed, somewhat competitively.

This was all very good. But then there was something more, for there was always something more on that *Beagle*, a mood of taking every small thing to bits. It happened when Capt's voice softened and he said for Covington to hear—when he pressed himself against the dark, crouching doglike in the companionway—that while the boy had a clever hand, and cunning wits, *there was something about him incorrigibly rough.*

He heard Earle protesting, 'I very much like him,' and another: 'Not at all, not at all,' though whether for or against, or who it was who spoke, he was not certain.

Covington marked FitzRoy down for his loveless snipes. He was a man who knew everyone between those timbers, down to the last cockroach riding the last scampering rat. He had a greedy eye for weather and all manner of terms for noting it down, so that every cloud had its own description. If the breeze, God send them one, raised the brim of a hat from the deck or tapped ever so gently in the rigging, it got a name and a number, as if it had appeared again, and was not a new minting of creation, but had only been out below the horizon awaiting its turn to be shewn to them once more like a shy bride returning.

One night Covington entered the mess with jam roly-poly, and they all four fell upon it with cries of sport in the nursery manner that spelled their liking for fun. Capt, gone on Madeira, cried in his piping voice, 'I had a black mammy who baked this sticky mess bedad I loved her!' Covington nipped back and told Pinchgut of his triumph, he should try this jam more often, and the cook clipped Covington on the ear, for it was every day that he baked it. Covington was new to his duties in the toffs' corner of the bark, he wailed, where they told all manner of tales to pass the time.

'What manner of tales?' Pinchgut sniffed advantage, holding back his hand. He was not attracted to rumours of how the world was made. He wanted a quiet time among his smoking coals and greasy griddles.

Covington smiled and licked his fingers, wiped them in his hair, said, 'Wouldn't you like to know,' and went and fetched his Polly Pochette. She had not been called for those many weeks of the voyage, but he would show his spirit a chance in an atmosphere that melted to butter.

They saw it, the winking curves of walnut wood. And presto Covington was enjoined to render a tune, a merry jig played in the inn near the crowded kennel where Spit and Polish were fart-daniels in his Pa's litter. Pelting over the bridge Covington bowed, raising a fine dust of resin. Soon his four were fox-hunting, with all the tally-hos and taran-taras in their tiny State Room, their sweaty shirts and stitch-busting breeches jerking around in the close air, the smells of their guts thickening the tropic night. Mr Earle went leapfrogging over the back of the gent with neither room to bend nor turn, and Capt deep in his cups was obliged to render Covington invisible to his emotion. He asked for his shoes to be wrenched off, and Covington hated doing it, but lo he fulfilled the task, the green insides slick with personal moisture. When Covington met his eye he looked startled. A scurry of shy brides again. It was better that Covington was not there. He went up on deck, out under the stars. The commotion continued under his feet; he could feel it through his flat foot-soles.

Any time such rumpus was heard from the gentry the crew took no notice, except to grumble, 'They should have a flogging for being so drunk,' and really it was as if their betters were farther away than the stars dusted through the cross-trees.

Covington was unlike that crew. When Covington found his betters at play he felt a longing to know the cause of it. He was one of those who conspire with their own fate, or else bewilder themselves to death. In this he was unlike the men. He was without the fatalism of the born sailor. And in this the crew sensed his difference, his longing for prefer-ment. Meantime the gentry repulsed him as being dissimilar to their kind absolutely, an upstart catechist with the instincts of a jack donkey.

Covington threaded his way to the bowsprit and hung over the oily waters, watching the figurehead in black reflection. The carpenter had him scrubbing the beagle-hound till her wood was white and hairy, and then painting her up. He hung over one hundred fathoms deep. He inked her eyes in the shining black that Mr Earle called *cachou de Laval*.

\*

They broke with the Doldrums and entered the Trades. Midshipman King got beside himself on the booms.

'We have a wind, now, Covington-chappy, a wind.'

'Aye, and thanks to ye,' thought Covington, 'I would not have felt it if ye had not said so!'

Now air bubbles twinkled in the *Beagle's* wake, ocean hissed against the side. Covington's wooden bitch went smiling as she angled her nose against the slippery deeps. Mr Earle could be seen sketching down the length of the deck, always at his pencils and keeping his hand in. Covington was doing a hard think. He had located more bumps on his head, smaller than split peas, cavities and corrugations. What did they mean, that he was wiser than he knew? For the Patagonians' heads were round as coconuts and they were foolish as horses. Likewise Volunteer Musters, his skull was simple as a plumseed.

MacCurdy saw Covington plucking his scalp and gave him a jab. 'There got he a *knock*, and down goes poor Cobby.'

He carried plates, cleaned up bones and gristle that had fallen to the deck. His watch was that of a dayman, an idler: by rights meaning he should sleep the night through, except he was called any hour to fix what was easy, running with a tar-bucket to stop leaks in the decking from wetting men's heads, and mending the seats that hung above the water on either side of the stem, that they called the Spice Islands, or heads, where they did their grunts.

A whistling came from the rigging as they heaved down a wave and counted the days till landfall. They made good speed. On such days they seemed to be motionless, or falling, yet FitzRoy crowded on sail. It was effortless, you might say, yet there was their commander on the quarter-deck with his legs planted firmly apart, his eyes narrowed and his gingery eyelashes watching every move the crew made. He spoke from the side of his mouth to Mr Wickham, who passed the orders along. At the far reach of his word men twitched every angle of sail and spread the vessel's wings like a bird.

Mr Wickham called Covington up and said, 'Make a good copy of the captain's weather notes, he wishes to see what you can do.'

'I don't find my captain "uncongenial",' thought Covington as he savoured the duty. 'I never had such thoughts and would never use such a word.' A grin fixed on his face till it hurt. Why, Capt had brought him righteous punishment just in time. The result was as he wished, his grief for Joey being calmed, making him merry and willing again. The Lord acted through Capt, God bless 'im.

Covington whistled as he dipped his pen and made his curlicues of d's and y's. FitzRoy loved all weather and had a lexicon for describing it. It seemed that FitzRoy had a great hero. It was William Dampier, in whose voyage to New Holland there was a mine of meteorology with respect to winds and storms. Covington saw Dampier, a buccaneer, somewhat like him, crusted with salt, burnt black at the eyelids, hard, observant, and as it were the devil's adversary in his law.

When all other duties were done, Covington cleaned. He whistled while he cleaned. Even the passengers cleaned.

'Open all hatches,' quoth Wickham, 'and let the ship air. Do your washing. Clean your bodies. We are not dirty in this navy, like Nick Frog's.' There would be the added matter of foul air and fevers once they reached their mooring, for which Bahia was famous. 'Let the ship smell sweet.'

They sailed through brown Brazilian waters, came about with a wallow, and entered the Bay of All Saints and saw the town of Bahia lying on the slopes and steaming after rain. The crew's jaws hung open at the idea of shoregoing but orders ran contrariwise. FitzRoy strode the quarterdeck, rattled his lid, and bade the sails snap in salute to other vessels, of which there were more than one hundred in the anchorage.

'Boy!' the cry went up. Covington grabbed the holystone and scampered along the deck. If Capt had his way Covington would spend his time in Brazil caulking and painting and winning Fleet prizes. At least he would not die of fever.

The foretopmen mocked Covington as they dangled across the sky: 'Pity the poor sailor, there will be no ran-tan-tara for Cobby.' They all had something on their minds to follow from getting their toes on terra firma. MacCurdy yelled he would be taking his pay in Portugee wine; Door bared his front teeth and said he would break curly heads; Robson, Hare and Rensfrey said they all ached for a handful of turf, and would grab it wherever, taking loud wagers on who would be first to harvest a moll. You would think from their bragging there was nothing else for it in Brazil but the breaking open of things.

Covington watched out for their gent, whose mind was

intent on getting ashore as strong as anyone else's. Darwin went below with King and Usborne, his designated companions for the day. Covington heard those midshipmen singing 'A-Hunting We Shall Go' and wondered, was he wanting too much to have what they had? What law under heaven said no?

On the water a black slave appeared. He poled a tree-trunk with his master in a rowboat alongside, being protected by a gaudy umbrella. Crew hung over the rail appreciating the picture, some of them in thought, while others laughed. A second party of blackies appeared poling a barge carrying a heavy weight of old iron; it was an anchor, and they were in grave danger of sinking. A third boatload was heard keening, and Musters opined it was a happy song, but as for Covington, his heart was made heavy observing such poor niggers at work, because they reminded him of one of the last matters he and Joey ever talked about.

'I doubt they are joyous,' said Covington, 'seeing as how they have cramp-rings on their toes.'

'*I* don't see no chains,' sniffed Musters, looking the other way.

'An' *I* don't give a damn for anyone,' added King, tossing a plumstone at the unfortunates. When none of them responded, King answered a question that was never posed: 'I have read all of Lord Byron, that's why.'

Nearby was a vessel of their own, the pride of the South American fleet, the man-of-war *Samarang* under Capt Paget, her sides as high as a golden cliff. Mr Wickham gave a great cheer, 'God bless old England!' and was answered by the roar of many throats. On their starboard side was a leaky old tub called the *Rio Trader*, a Carolina slaver of ill-repute. Some of the men spat to see her, while others gave the low hurrah as men do when they take sides in an argument. On board the *Beagle* it was often said that slavery was a good thing in the hot countries; how families who were against slavery at home, such as the Darwins of Derbyshire, would soon change their tune in tropic parts, where slaves could be heard singing at the tops of their voices. Chief of all slave enthusiasts was Volunteer Musters, a very schoolboy in all his virtues, full of facts about men expressed in a voice that belonged in choirs. '*I would have one*,' says he, 'I would have six or seven of the rogues.' Covington gave allowance to Musters as one who had never dirtied his hammock with longing, who knew his fellow-men only as he knew toys. It was in this resemblance to Joey Middleton that Covington liked him; though in his difference from Joey—a want of heart—he itched to set him right.

Bahia made a fine perch for Portuguese men of wealth, with its whitewashed walls of mansions and churches, and palm trees clutching the sky. So bright was the sun that the walls created squared-off shadows in the afterlight of men's stares. The slave-masters lived above the swamps, wherein they tipped their chamber pots and flushed their drains, and if one of their Africans died by breathing the miasma, why then, it little mattered, they would go to the markets and bargain themselves another. Phipps ranted against the practice, speaking from the corner of his mouth as if he would strangle each word. Bahia was ever a place of slavery, he said, where a fellow for a few vintem could have another being for his keeping, and every Man Jack could have for his Miss a chosen black beauty—indeed, as many as his pocket and their jealousy allowed. Until recent times the murderous Portuguese ordered themselves carried in hammocks through the streets, pausing where it pleased them, making conversation while their slaves hung the hammocks on metal spikes, bearing their weight under the blazing sun of twelve degrees south.

Gent with his gear hauled himself on deck and awaited rowing to shore. Capt and the gent kept a distance from each other as was proper in those busy hours. Gent's tools were a rock hammer and a gun. Under his arm he clutched a spyglass, that the Fuegians with good naval flair called the 'bring 'em near'. He swept the wooded hillsides with it, capping and uncapping it excitedly, grabbing it to his eye while sweat tickled his cheeks. On his back he carried his rectangular basket slung by a canvas strap. He said he wished to enter the forests and clamber the hills and get himself a paroquet as soon as he could, and Covington caught the coat-tails of a dispute in his peevishness; namely, that Capt would not allow him a boat just yet.

During their waiting little Musters sped at Mr D, pretending to wield his Do-Little Sword, and the gent got his spirits up and played along, crying '*Allons!*' and praising Musters's swordplay as he slumped against the mainmast: '*Un petit morceau de tout droit, monsieur, au revoir.*'

Covington turned his back on them, tears pricking his eyes. He had the feeling of the boy who shambles through the schoolyard alone, too old for play, too young for mastery of his fellows; or mastery of himself, either, if the truth be told: the one left out of company who plots his victories thro' jealousy. He hoped Capt would relent his passions and give him leave to range about with his pocket-violin, as when he played in parks and along seawalls in Monte Video del Mar under Capt King. Covington appealed to Capt FitzRoy with a damp eye. Capt saw him not.

'Tar bucket!' called Mr Wickham.


Covington worked around the foremast on his knees, caulking away to the strains of squeaky music. It was Carnival time ashore and the noise of it crossed the water and tickled the ear. He wanted to be gone from his duties, clicking his heels on cobblestones. He saw that Miss Basket was in the same mood. In her shoregoing finery she timed a jig with her rump and leaned over the rail, where her eyes tangled with Covington's. You could almost say she chanced a smile; or as close to a smile as those gloomy Patagonians allowed. Down the back of her neck hung her 'follow-me-lads', curls and ribbons gracing a plump shoulder. 'Whoa!' Covington cried, in a show of high spirits, while Revd Matthews warned him with a hand gesture to bear away. 'Boo!' Covington told him, and Fuegia poked out her tongue. Matthews said, 'I w-warn you, Master Covington. I mean to reach Chapel without splashes of tar on her p-p-person.' So Covington set the tar bucket down, and Yorker being nowhere around, and Capt barking orders over the side, and Matthews thus blathering, he grabbed Fuegia's elbow and juggled her into his arms, whirling her behind the yawl amidships.

Matthews tapped his shoulder trepidatiously as he let her go: 'I say, Master Covington, what's this?'

'Wouldn't ye like to know.'

It was not until smoky sunset the next day that Covington won his ship's leave to go ashore. Mr Earle sat in their boat with his knapsack of paints. Door manned the sweep oar and contrived to bang Covington's shoulder with every pull. Covington wore white dimity trowsers fringed at the bottom, a muslin blouse with loose folds wherein he wrapped his Polly Pochette, and low on his eyes a straw hat bargained from John Phipps. He was easy and independent in his humour.

When Mr Earle turned his grin on Covington and asked if he would carry his load when they got to shore, Covington answered, 'Carry your load yourself, sir,' and Earle much liked him for his cheek. 'Anyways,' Covington added confidingly, 'there are little boys for that purpose on the dockside.' Then, more loudly, he boasted past knowledge of Bahia's putrid alleyways, inventing jolly inns and pockmarked beauties standing in every *Door*. Mr Earle choked: 'I hope you have brought presents for all your little bastards, Syms Covington,' thus making the lads howl.

'I keep no count of my by-blows,' Covington answered, with Door knocking him sideways again. Thus his good spirits made play. Door stood in the stern, monument to a sailor's vanity, making Covington full proud to know him as one of their bark's finest; wearing a scarlet waistcoat tied with black ribbon, white dimity trowsers ditto unto

Covington's, his black hair oiled and tied in a pigtail, and, kept handy, a smart switch fashioned from the backbone of a shark.

Door leaned at Covington, and sang, 'Be cheery, dear Cobby, let your heart never fail, while the bold *harpooneer* is striking the *whale*.'

Aye, Door was a rare lad. He stood ready, at last, to admit Covington into the charmed company of devils, which name Covington gave to the foretopmen ashore.

The *Samarang*'s lads spotted them from the man-of-war's high decks. The *Beagle*'s boat replied in like spirit, 'Huzzah!' as they dipped under her bows. From aboard that great vessel came a buzz of activity that was like a beehive; it made Covington feel he lived in a walnut shell with hydrographic survey poor cousin to battle orders. But even so, England's navy ruled the waves. Light shone from the *Samarang*'s gunports and tickled the water, mixing the colour of honey. Wee Volunteer Musters was drawn to the scene most keenly, the glory of battle as fought in Nelson's time being his greatest hope.

As they pulled across the bay Mr Earle reclined easy as an admiral, and continued philosophical: 'Why,' he spouted, 'the children of Love are more naturally and properly the heirs of a man's inheritance than the unwished-for consequence of dull conjugal duty.' Hrrumph to that, noted Covington, knowing it was Earle's way of saying he would spike a maid ere midnight, and damn the consequences.

As they docked Covington asked Musters if he was a-coming with him, but Musters bade him curt farewell in his piping voice, and leapt into the *Samarang*'s boat among other boys, and was forthwith taken over to mingle with his gods.

'Where to now?' the shipmates asked, jostling each other, singing, '*Once before we fill, and once before we light*,' and telling each other, 'Long may your big jib draw,' and other low boasts about the night they had coming upon them, when they would get themselves drunk and dance the matrimonial polka. They knocked Covington around the head and called him a dog's pizzle, but he allowed them to do as they wished, even to lifting him up and tossing him in the air while Augustus Earle paced alongside making his gap-toothed grin. They linked arms and hauled through crowds that were busy tossing bladders filled with water, which hit them, not Covington, putting the lads and Earle in a foul temper with all natives. It was there, between the waterfront and the upper town, that Covington contrived to lose them at a wrong turn, and found himself alone with his Polly. 'Seek and thou shalt find,' said the scriptures. So Covington set forth on the next adventure of his life, from which there would be no turning back.

**H**e was not alone, exactly, for this was a night when slaves ran free all over the town. As Covington slipped between hosts of shadows he was grabbed by strangers whose fingers traced the lumpy ridge of his nose and whose thumbs dented the dimple of his chin in wondrous regard. Once, easy friendship would have grown from such prods. But Covington had no patience that evening. None either for the pleasures advertised dog-cheap at the door of every cunney-warren, where maids stood '*in cuerpo*' as was said, with their gowns falling open. Any foreign sailor they enticed was called John, and so Covington was called John Corona, meaning Crown, for his big pleasant head.

He lowered his chin and passed into the night. The air was thick as smoke. Lightning lit the heavens. At the far back of the town (high above a wild forest, as he was to learn) he was sent cowering into a stone doorway by a great rainstorm. The drops were big as silver eggs, breaking with a splash. There was nowhere else to go. Covington heard rapid guitar sounds, and found that where he stood was entrance to a donkey stable. Within, a quick-time was made that dinned his ears, out-sounding the deluge and setting his pulse racing.

The place was lit by lanterns. All around, some standing, some with legs up on rough benches, were gangs of mulatto youths wearing bandannas, loose shirts and piratical pantaloons. They bade Covington welcome, and one he saw played a small four-stringed guitar like a ukulele, and one rattled a hand drum, and one tapped with a stick, and one had the face of a pug-dog: and that was the one who sang; and she was the ugliest wench Covington ever saw; but he must have been bewitched, because the voice that welled from her made his chest swell and his eyes water as if he knew all love in that instant, and would every time she groaned her passion. She was a dwarf who minted gold with every hard note she pushed with her breath, and between rounds drank from a goatskin of *vinho sangorino*. She said that her name was Leza, meaning beauty, a word Covington knew from *belleza* in the Español; and cocked her nose in the air with proud and ridiculous defiance as she spoke, for a beauty she was not, and never had been. Her ankles were thick and her bare feet stood flat on the floor like a hippopotamus's, and her thumbs were double-jointed and very long-nailed.

Covington experienced a longing to dance, to forget his life in its former part, and give welcome to the next, yet he soon found, if he dared essay a step with any but her, that this Leza came seeking him out, calling him John Corona—and grumbling at him as if he had done her special injury. She drew him to her side as if she would own him; and he swore from the way she looked at him that her songs had grown to be about him.

Covington did not like her game. It made him combative. One of the youths began to glare at him and mutter harm. It seemed Covington had displaced a favourite, the boy who played the *cava-quin-ho*, which name they gave to their wailing ukulele.

Some time in the night Covington pulled his Polly Pochette from under a bench and played English airs. The gang laughed at his music and Leza spat. He found resin for his bow. He blew up a storm of notes to tempt their regard, and still they laughed, and rudely talked among themselves. But in a short while Covington hit upon a tender note; Leza made response with her vocal cords and sounded it too; it seemed they made it together; she rolled an eye, her throat shook like a string, and Covington's Polly pierced the gloom with a note she had never made before. Here was a strange coupling for you, as in like manner they played one instrument, Her Ugliness with her voice, Covington with his Polly-Meow. Dancers went past in close embrace; Leza moaned; there were shadows against the walls, and the *vinho* went round more times than Covington remembered. Then he knew not who he was, nor where, nor what hour it was.

Except that when, with a clatter, his Polly fell loose from his arms, he knew himself drunk as a rolling fart. He had had his fill of the fetid air and wanted to clear his nostrils, but was he allowed to leave? Ninepence to nothing he was. The singing showed him love, both sides of the coin. Leza had a grip on his waist like a bear's, and a breath as bad. Covington had scant Portuguese but understood the words of her songs too well: they spoke of jealousy and rage, and *he* was her chosen one, and *he* would have none of it, and so *he* gave her a kick in the haunch that daresay inflamed her passion to hatred. 'Lua na testy munha,' she sang, which meant the moon was her only companion.

Covington saw the glint of a cutlass and was flung from the donkey palace, marched to the head of a ravine and bucketed in a tremendous roaring rainstorm. From there he was given a push. A pit opened under him. Struck a farewell blow of metal he went sliding down a muddy slope. Broken-toothed laughter followed, then faded.

For now, the town of Bahia was gone, swallowed into the no-place above. A bolt of lightning hissed. Covington's legs jerked and his arms flailed, and he swallowed much mud. He tumbled one hundred feet.

To his surprise, when he stopped rolling, he heard small birds twittering in a forest, and saw through the mist the greyness of morning.

C ovington had been struck on the forehead and bled a trickle. His pocket violin had her bridge broken and strings awry, but was otherwise intact. He huddled her to him and promised repair. The rain stopped and cloud lifted away from the trees in misty patches. The sun came over a rise and smacked him in the eyes. 'Hey ho,' he whistled, 'what am I to do?' *Laugh*, for what else was there to do in the pit of fortune?

He strung his breeches on a bush and waited for them to dry. He bathed in a nearby stream that amazed him with its many clam shells, all pearly within. Spotted flowers were everywhere on the forest floor, and ropes of vines. He could not have been distant from the Beagle but had never been so far inland on any of his earlier voyaging. It felt like another world, a better and a stranger one. His Polly sounded dull and damp when he plunked her in his lap, but Lord, she had known worse. Covington fought battle with a giant mosquito, and won.

He craned his neck. Bahia stood as distant from his hopes as poop deck stood from topgallant forecastle, or a boneyard stood from the House of Lords. But would that ever discourage him? Fixing his eyes upwards he gave thanks for life, and saw through a gap in the trees his fall-marks streaked from ledge to ledge. There was no trail back the way he had come. When he tried to re-ascend he slithered down again.

He was late returning and knew he was up for a flogging. But what matter, he thought, still somewhat dizzy, 'Let me conspire with fate or die in a devil's nightcap.' He returned to the stream and cooled himself in the shallows. His head cleared. He put a feather in his Phipps's hat, that had half-melted in the rain, and with a strip of vine wove a necklace of small flowers. He was filled with a bubbling bravado, finding joy in everything, even to a leech that sucked his ankle of bad blood and swelled to the size of a chestnut. He cleaned his fingernails with a split shell that he tucked in his waistband, took up his broken Polly and wandered a muddy, meandering path through the deeps of the forest. He gave thought to the idea that Brazil might become his home in preference to a flogging.

Thus dreaming of estates where he would keep slaves, but in kindly comfort, selecting females for his use, Covington went on for a time; when, rounding a bend, he heard loud gunshot echoing through trees, and winced, thinking it was aimed in his direction. But then through the woods he saw a figure coming, and lo—and to a great laughter welling inside him—it resolved itself into the spectacle of Mr Midshipman King, jaunty as a fresh cheese.

Covington ducked down to hide himself from an unwanted encounter and planned to stay there, hunched as a stoat, when a pigeon flapped at his feet, stunned but otherwise uninjured. Covington's spirit overcame him. He gathered the creature into his hands, sprang up full square to show himself in a column of hazy sunlight, making himself, as it were, into a vision of the kind Mr Earle delighted rendering in water paints—the English sailor ashore in a forest clearing, clad somewhat in the native fashion.

King spluttered, 'Covington? What in the name of Zeus?!'

Covington could not restrain himself, but cackled, and threw the bird up to deny King his prey. As the bird gave a few determined flaps King at close range raised his blunderbuss and let fly with fearsome accuracy. Powder stung Covington's nostrils and the report made a ringing in his ears. Of the bird there was little more to tell: when the smoke cleared there were only a few floating feathers.

King congratulated himself as a great huntsman might, by pounding his chest. 'Thank you, Covington, what sport! Say we taught that bird to fly!'

Only then was Covington aware of a second figure traipsing from the side bushes with an armful of greenery. King pointed at Covington as he scrambled to his feet, 'Boom-boom!' and their gent—for it was he—greeted Covington oddly:

'*Levar flor! Levar flor!*' he muttered, very red in the face. Covington was dumb-foozled. Here was the man who had fingered Covington's scalp; who saw Covington every day within the small compass of their walnut shell; but who mistook Covington for a peon of the place.

'*¿Que?*' Covington responded, a vicious repartee in his head.

'*Flor ... levar flor ...* flowers ... carry to *pueblo* ... to town ... '

'Darwin, you're a nincompoop,' chortled King. 'Look what he carries—a fiddle!'

The gent laid his seed pods and branchlets on the ground in a tumble of green. 'Too much, too many,' he grunted, then straightened and met Covington's eye. '*You?*'

Covington raised his hat: 'Aye.'

'Ah, yes, how could I forget?' he puffed, his phrasing as delicate as if he had said *lumpy noggin*, and then: 'What brings you to the forest, sailor?'

'*Alone*,' interposed King—it being against Capt's orders to venture ashore without companions in foreign ports.

'I wish I could easily tell you,' Covington answered. 'For I was set upon by a gang and pushed over a cliff.'

'Is it such a dangerous town? I had not thought so.'

'Covington is a rogue,' said King. 'You must not believe a word he tells you.'

Covington bowed, crimping his annoyance. 'And you must believe your ha'penny is good silver,' he thought to himself.

Darwin studied Covington and Covington studied Darwin in return. This was in effect their first meeting. It would ever be Covington's pride to cloak dependence on good opinion by sending his glance back unflickering.

'He's cut,' said King.

'I am sober,' said Covington.

'You have a nasty *wound*,' said Darwin, and fingered Covington's temple, finding a tender place where skin peeled open. 'I loathe to do this,' he added aside to King as if Covington had not jug-ears to listen, and then performed a rough bandaging on a gouge that had not bothered Covington at all, using a roll of cotton or tow kept in his rectangular basket that bore the sickly smell of putrefaction.

Covington spluttered his thanks: 'Sir ... '

'It is nothing,' said he. 'Now find your way to the ship, taking care not to stumble.'

Words stuck in Covington's throat. He wanted volumes to speak out what was in him. He knew in his heart it boiled down to a plea: *Know me.*

'Mr D?'

Darwin swung back on Covington, smiling to hear a nickname coined so unaffectedly. 'Follow the pathway, take the fork by the mulatto's farm, where they keep spotted pigs and grow mango trees. Understand what I say?'

'*Sir*,' Covington almost pleaded, still trying to get something out that choked him and surprised him by its occurrence in his heart.

'Yes?' with puzzled impatience. 'What else, lad?'

Must Covington remind the young gent of what he'd found in him by kneading his nut by the ship's rail? *Friendliness, helpfulness, adhesiveness, amativeness,* just to name four prominences that might yet determine the direction of Covington's life.

'I am well,' Covington said.

'Yes, Sailor Covington, excellent, you are well. You are bared to the bone by a cutlass, but you are well. Indeed you are well. Now along with you!'

Covington saw he must fling himself home to the gent if there was any hope in heaven.

'I cannot,' Covington said.

'Cannot?'

'Meaning will not,' quipped King, making a sign with his hands for Darwin's eyes, slashing the fingers of his right hand across the palm of his left, representing a man being flogged.

'Oh, my Lord, I had not thought of *that*,' Darwin sighed, and sank down onto a rotting log that shifted slightly under his weight, sending out a scurry of beetles that he followed with a glazed eye. His moleskin breeches were stained and caked with dirt; his skin was lumpy with insect bites; beads of sweat decorated his upper lip. Covington liked the picture he made, it was young and adventuresome, beginning a time of hero-worship. But he was half-gone with exhaustion and drink, and impatient, and without further talk scooped his belongings from the ground and stood ready to attend the gent's needs, leaving him only his birdbasket, that was strapped to his back.

'What now?' the gent growled, submitting and objecting simultaneously to Covington's proffered arrangements.

'Covington is our hammerman!' shrilled King.

'Indeed?'

Covington half-bowed, acknowledging his place on earth as a servant. At the same time he was quick to make it seem as if he had been done an injury, or would be, if his arrangements were questioned. This pertness in his adopting a role unasked-for put Darwin on his mettle, and showed Covington his way forward with him. If a medal was stamped to show their relationship henceforward, it would have sunrays leading back to this moment.

Darwin and King began walking and Covington followed. When they slithered down a slope of red soil and wandered into a country of ferns lofty as a cathedral Covington went along. At every turn Covington came on a little, and Darwin, easing the discomfort Covington's presence caused him, chose to beg assistance. 'Come over here.'

'Squire?' Covington leapt forward.

'Are you afraid of snakes?'

'I have yet to feel it,' Covington boasted.

'*Nota bene*,' said Darwin, with his boot planted on something threshing in the leaves. Covington leapt back. The other laughed: 'What do you call that?'

Covington peered close. 'A serpent, sir.' But the creature had legs, and Darwin laughed again, proving Covington's ignorance in the natural history of dry land. The creature was as big as a blanket-roll with teeth like a trap, and a head raised in proud alarm. It ran towards King, who leapt and yelled. It changed direction and rollicked towards Darwin, who made a gleeful cry, raised his gun, and blasted it front-on.

'Do you call this duty?' cried King.

'Nay, sheer delight,' Darwin answered him.

Soon Covington found that, as well as his Polly Pochette, he carried seed pods on a string around his waist, and a crude bag woven of banana leaves, in which he hefted a weight of colourful clam shells from the stream bed he had washed in earlier. Still he was willing to carry more! Still the young naturalist kept scooping things up!

The noonday sun was high when King sidled up to Covington and said, 'We are suffering a great heat.'

Covington said nothing.

'He believes you are in for a flogging, Covington.'

Covington stared back at King wide-eyed, as if to say, we have not spake of the cat!

'*He* is your only chance,' King persisted. 'You shan't have a hope with me.'

'Hush!' from up ahead.

Darwin raised a warning hand. On a low branch huddled a bird with a long curved beak. Its tail was flat as a paddle, its feathers black as *cachou de Laval*. It had a pale throat-patch that hung in the darkness as pretty as any white peach of Kent.

Darwin took the gun from King and loaded a smaller kind of shot. From branch to branch in the shadows the bird stepped along, squawking low. Covington squatted on his heels and watched. He had scant knowledge of shooting yet admired the hunting mode displayed, the naturalist not spreading himself cheesily in the woods like King, but turning thin sidelong to his target despite his large frame, and raising the gun-barrel deceptively like something in nature. When he fired, the bird fell among brambles with a thud.

'There's a pickle,' he muttered, and stood with a hand to his belt.

'Poor sport, Darwin,' carped the Midshipman. 'Our kingdom for a dog?'

Darwin angled his head, and smiled faintly perspiring at Covington plaintively staring from his ground. Covington beamed his wishes and they lit on the young master with unspoken understanding. *Should I yap?* Darwin gave the barest nod.

Slipping his load from his shoulders Covington plunged into the thicket. Angry buzzing flies stung him. He went on hands and knees. The soothing gent's voice coaxed him: 'Be careful, there. Go a little to your left. Mind your bandage. Good man, fine fellow!' Covington crawled in a torment of existence. Welts came on his skin from a hairy vine. Gnats stung him. But he would not have wished himself anywhere else on earth.

He found the bird under his nose already teeming with ants, as if it were made of sugar. Shuffling backwards he emerged holding the dumpy creature unmarked by gunshot, except it oozed blood, as Covington did himself from under his bandage.

'Whizz-bang, jolly good!' exclaimed King.

Darwin brushed the ants from the feathers and pronounced the bird as belonging to the family of toucans, notable for their great beaks disproportionate to the rest of their bodies. He addressed this information to King, in the manner of King himself—making all things obvious— Covington's brain noting throughout that his name was not used, through having been forgotten again, or withheld in discouragement to over-familiarity, or because with servants it was always just so, when all he longed for was acknowledgement, say as a pet desires a fondle around the ears, as the marmoset Joey had been to his handlers.

While Darwin stood by, King it was who took a little cotton and placed it in the bill and nostrils of the bird to stem its bleeding; King who took a pocket-knife in one hand and with the other parted the feathers along the bird's breastbone, making a rough cut between the outer and inner skin, while Covington knew enough of careful work to know it was damned clumsy.

Darwin growled: 'Recall, as I showed you, King, use

patience, use care ... ' (King went slow.) 'Yet do the whole thing promptly, as I say.' (King went fast.)

A memory returned to Covington's brain. In the first weeks when he never knew this gent except as a piece of yellow bile on a letter-case, King was given his leave to go roaming with a gun. Covington gave thought to who King's companion was on those countryside rambles. Covington's unconscionable jealousy came in a whirl as he stood in the steamy forest, breathing hard, observing the task at hand.

He followed how King treated the head of that bird, for he meant to do better when his day came. King snicked the skin of the ears and tore it over. The eyes soon appeared, as if from the inside of a glove; one broke and the other had a covering of filmy skin that Covington shivered to see, because it still seemed to be looking. King went on fiercely in a panic until the gent hooked him by the shoulder, steered him aside, and extracted the other eye using the point of a quill. 'Botched,' he said, stifling extreme irritation. King must have thought he said 'Tops,' for he grinned showing his long teeth and batted his morbid eyelids. Darwin in a few minutes had the feathered cape of the bird smoothed on a sheet of paper.

'There so,' and the dripping, naked corpse lay in the grass. King whistled his achievement. Gent examined the flesh in silent fashion, and said he would cook it and keep the bones. Everything went into the basket, which Covington carried now, and they set forth to escape the forest.

Covington went ahead of the other two, pretending he was no barnacle on Gent's hull, but really ensuring there was no fork in the track where they might lose each other. If he came to one he looked back and Gent nodded. Within the half hour he found himself at the edge of the lower town, where a thatched hostelry offered meals and grog. Smartly he made himself understood to a Negro, and when Darwin and King arrived was heard calling for 'a cup of the good Sangorino for the officers, pray'. He had their beakers ready. He had other matters in hand, including the lighting of a fire and the procurement of a cookpot. Ere long the bird was boiled and cleaned of its flesh, and Gent thanked Covington for his trouble—even to taking a taste of the meat and pronouncing it fair, and bidding Covington to suck upon wine at his expense, and so Covington did, but after his previous night's rort a single cup made him ill in a patch of banana trees; meantime, Darwin and King made merry as schoolboys.

Covington sat shivering on a bench until they were ready, and it was time to follow them to the harbour. He had the beginnings of a fever. The truth was he had never been entirely well since his flogging. From the dockside the cutter was signalled to collect them. Door manned the sweep oar, raising his eyebrows with bemused expectation on sighting Covington all tattered, licking his lips and muttering into his face, 'Well oh well, 'ere's some Christmas beef all fit for slicin'.'

Climbing the ship's ladder Covington lifted his eyes and saw Capt's lamp burning in his cabin. So he was home. Indeed he was home. So he was doing his tally of miscreants. Indeed he was adding 'em up. Covington felt a trembling in his shanks and cursed himself for cowardice as he traipsed along deck. John Phipps's eyes were upon him, somewhat cold. He had been gone from their bark near twenty-four hours. He would be hanged if Capt read the laws true. And when a hideous Capt emerged from below, and leered, it seemed he would be: 'I am putting ye to the hoop,' lipped a horned figure with burning eyeballs and the tongue of a lizard. The whole catalogue of a commander's crimes came to him in the grin of a jack-a-dandy—for it was Jemmy Button, a Fuegian wanting clams for his chowder, and not the captain at all. *I am putting ye in my soup.* In hand-me-down gloves and polished shoes Jemmy was a right proper hog in armour. 'I have mistook a savage for an aristocrat,' grunted Covington. Jemmy's English was contained to a few words. 'Where you bin gone, Sailor Cobby?'

'*I* have "bin gone" nowhere, old sport, 'cept under King's Orders,' Covington replied, holding his breath and getting jigged up and down with Jemmy's arm around him.

'Sailor Covington?' The voice was Darwin's.

'Aye?'

With the gent using his name Covington's blood was refreshed.

'Follow me.'

'*Aye.*'

Ask Covington what happened next and he would burst his gizzard to tell of it. By silent consent of Capt he accompanied Darwin to his cabin—the cramped poop cabin, which he shared with two others—where he found himself occupied in cleaning, drying, storing and, if ye will believe it, in dispersing the rankest shellfish to the Fuegians' kettle. The gent solicitously asked, 'Are you ill? Are you up to this?' and when Covington said nay and yea, he disbelieved him, and gave him a swig of gripe water, and a few drops of opiate, after which Covington was fit to carry on in true naval fashion, shedding all tiredness and standing to his colours.

If Covington looked like a beggar in his torn trowsers then so did his gent. There began with him the idea of brothers, the good elder and the willing younger. There began so many other ideas that his brain was a bucket of jewels to think of them.

Covington whistled at his work, meriting more praise than he earned in dreams. He copied Darwin's way of prising open shells with a knife, cleaning them of flesh, washing them out, and tying them together with a piece of thread while keeping the hinge intact. The smaller shells the naturalist told him to pack inside the larger, to save room, and to ensure their safety. The cleaning they completed on deck until rain drove them inside. Covington was told that a scientific collector in England would give more for a shell covered by its rough coating than when it had been taken off by unskilful hands.

'I shall keep that in my mind,' he said, with a forwardness that prepared his way.

With Darwin that long day Covington experienced what he knew, but hardly dared hope, would free him from punishment this time round: namely, that any sailor who assisted the ship's gent was given goodwill by certain officers for the reason that the gent was close with Capt, their avenging Moses, whose law was writ in officers' commissions.

As the work went along Covington heard these words: 'Good man, well done, you have been a brick,' and other such let-outs of breath as occurred to the one he chose and would serve unto eternity if he was wanted. They settled snug in the poop *cabaña* where Darwin let go the preposterous strain gents had about them, even between themselves. He culminated in examining Covington's thoughts as he expressed them, rather than by fossicking around the root-hairs of his nut and making suppositions. The boy liked this better, and his vanity swelled. It came about that Covington knocked shells to the floor that were arranged for sketching, and was able to set them from memory back just as they were, making piles of uninjured, cracked, small, large and curious, all ready for the gent's pencil.

Darwin turned from the table.

'This is good, Covington.'

'Yea?'

'The captain praises your wits. *I* praise your eye.'

'I look *and* listen,' Covington replied. Then rubbed his fishy hands through his hair and gave out a grin. 'That is not all the captain says about me.'

'I would not know,' said Darwin, his jaw tightening a little. There would never be any gossip from him, and that was that.

'He is a good Christian, our Capt,' said Covington.

'Indeed he is.'

'I am a believer too, as the Lord is my witness.'

'I can see that about you.'

'Can you indeed?' Covington was interested.

Darwin looked uncomfortable. 'It is a manner of speaking.'

'You have seen me at my texts?' insisted the boy. He was deaf to any sarcasm that Darwin allowed himself.

At that moment, seeing a barrel of dirty salt that needed moving, Covington enfolded his arms around it, and raised it to a higher shelf.

'Not there,' Darwin said. 'In the hold if you please. With my other stores.'

Covington staggered below, where the air was fetid and always smelled of mildew and dead rats, and of a fishes' graveyard. He stowed the salt between the parlour furniture of the Fuegians and a few labelled jars, and returned wiping his hands in quick-time down the side of his breeches.

'You have good strength,' said Darwin.

'I have, if I am given good work,' Covington boasted. 'Lifting, carrying, hauling. It is all the same to me.'

'You have concentrativeness, too.'

Covington laughed, gladly remembering that great word and a few others besides.

'Where are you from?' Darwin asked.

'The topgallant forecastle,' Covington quipped, 'above the coalhole. That's where I swing my hammock. I sometimes believe I was born there, 'cause whenever I want better for myself, I get sent back there.'

'Your accent is Bedfordshire,' said Darwin.

'Just so. There is no room for me there any more,' Covington nodded, 'except in the open fields. But then our John Phipps gave me a taste of the sea. Us Covingtons go back before Oliver Cromwell's time, God bless 'im.'

'Your people are dissenters, still?'

'They are butchers and horse cappers,' Covington nodded, 'that is what they are. They love their meeting house and their ale.'

'Bone men too?'

'Aye. By the drayload.'

'I am always chasing bones.'

'It is a long road round, to go back to Bedford for them.'

'Agreed, we are better off in America.'

'We?' said Covington with dim hunger in his voice.

'Tell me more about you,' said Darwin.

'They say I was born in a stable and wrapped in a horse-blanket. I do remember this—on cold nights we put our feet in warm horse-cack. That is barely all I remember till I was taken to chapel, and then I saw a young man leaping a stile. He was in a stained-glass window. I kept reaching out to it.'

'You wanted to break the glass?'

'Nay, but I am telling you something curious. That young man had a round face and light-coloured hair, which puts me in mind of you, sir.'

'Indeed,' said Darwin a little remotely.

Covington leaned his hands on the table. The day had worn him down. He'd had nary a wink of sleep and his fever tackled him. Darwin peered at him close. 'Use a chair, if you please,' and sat Covington down. He felt pinched at the knees from the tight fit. He had legs like callipers and bumpety wrists that slung from his sleeves like bollards. And so did his brother Darwin have long legs, bumpety wrists, and so on.

'It must be no great joy,' said Covington, 'to be wedged in here writin' out your thoughts, while the sea bucks you around, and Lieutenant Stokes draws his charts at the same time, aye Mr Darwin?'

'It has its compensations,' said the gent, and Covington believed from the light in his steady eyes that he meant entertaining Covington himself in that steamy tinderbox might be one of them. 'Covington,' he stared at the boy roundly, 'you are very ill. Don't deceive me with it.'

In his guts, Covington feared, he was working up to a run for the Spice Islands. But he shook his head. He didn't want to leave where he was. 'I am fit as a pudding in a dog's mouth,' he boasted as he looked around him. He had been in and out the poop cabin carrying cocoa, tea, beef on a platter and pease porridge by the bucketful; and looking in those books, that one time with Mr Stebbings; but had never looked around him as he did then, with such ease and freedom.

'It is a real snug little home you have in here. You must be warm as a mouse in a churn.'

'When it is calm I like it better than anywhere else in the world,' acknowledged Darwin.

'Is that where you place your pillow?' Covington gestured at a plank.

'Yea, and I can never get my legs fully stretched.'

'It is easy to see why not—if you never make space behind your head.'

'But then I would have no place for my pistol box.'

'Your pistol box might stand on its end.'

'True but where?'

'In that corner of the bookshelf that is empty.'

'Ah,' said the gent, 'but that bookshelf is the preserve of Mr Stebbings' library. There are books that fit in that place.'

'Where are those books now?' asked Covington.

'All around the ship.'

'And will they be always around the ship?'

'No. They come back each day. And so the space is filled.'

'But some others will always be a-going out?'

'Aye.'

'Then there will always be a hole ready,' said Covington, 'and another just the right size for you to fit your head in. It would not do for me, but your head is not such a great enlargement as mine.'

'The better for thinking, say I,' said the gent.

'And whatever else bumpology wants of it, eh?'

'Bumpology—indeed.'

'It is not thinking that I do with my knob,' replied Covington. 'What is the point? Us Covingtons are not made for thinking. Ask me and I shall do a job for you, though. Like this board that gets in your way. It can be done quick smart with just a hammer and awl.'

'Perhaps, perhaps.'

The gent was getting too much of Covington altogether. Sometimes there was never enough air in a ship even for one.

Covington did not wish to go.

His hand strayed to a stone ink-jar on the table and he closed his fist around it, feeling the coolness inside. 'I want to show you my hand,' he said.

'Let me see your copperplate another time. I have seen a lot today.'

Covington felt a darkening inside him, prelude to a faint, and told himself to force the day to its very end, for he had that power, to unfold his advantages and take his chances to the peak of his blood.

So he dipped a quill, and pulling over a sheet of paper he duplicated a piece of writing he saw in front of him. What it was, and whether Darwin continued his praise thereafter, Covington never knew, because at this pass he fainted clean away.

His head hit the table. He made a sticky mess for all to see. Mr Wickham came in, and Capt came in, and they called for Ash, the gunroom steward, and Fuller, Capt's steward; and those two oxen carried Covington by the arms and legs to the opposite end of the bark, where he was draped in his hammock above the coalhole and left to sweat. Darwin did not visit him, but sent a steward with eucalyptus drops to freshen Covington's air, and dose him with more opiate, which peopled his dreams.

When Covington came to his senses it was morning and Phipps was there to visit him, peering into his face so closely that the first thing Covington saw was a small mite creeping through the tangles of the catechist's curly black beard.

'Cobby old cuff,' he felt his shoulder being shaken.

'Where have you sprung from, John Crow?'

'Right from where I sit, here on this chest. I have recited half the gospels to you and you gave me your responses all over the place. You had the Temple Mount at Sinai and the Israelites were never in the Red Sea at all, 'cause you had them in the Dead Sea getting baptised.'

Covington told him his surprise that it was already the next day. 'It must have been my ghost doing all that talk.'

'You are a great knock, Rosin-the-bow. It is the third day since they put you here, and you have been raging like a hurricane and sweatin' like a river.'

'I am as cool as you like,' Covington asserted, and swung to the deck. But when he touched the damp timbers he collapsed in the legs. Phipps hoisted him back up and told him, in answer to a question Covington posed him, that there had been no punishments listed; that as far he knew Covington was Gent's missy now, because Gent had speculated to Lieutenant Rowlett that he might send for an order of trowsers for him, to be cut from cotton duck, to replace the ones that had been torn by Covington in the forest.

'You are hurt by this, John?'

'Hurt? Nay. You take me for a fool. I have fallen to the deck in a foul wind from a height of fifty feet and broken several bones. That is what I call hurt.'

'Good, then.'

'This other is not hurt, young sinner.'

'Do you have a fear of what I want, then?'

'No, for if I had I would have expressed it before now—because you have been going around this bark like a lovesick seagull with your eyes on that gent for the best part of our whole voyage, and what have I said about it before? Nothing, as I recall.'

'Nay, but your eyes have condemned me sometimes.'

'Eyes are but mirrors betimes.'

'Very well. But will you say what makes you so sour?'

'Well, I can hardly say. Maybe it is that I have come here to watch over you, and that gent, who you brought so close to your desires, has neither sent nor asked nor given you a half farthing of his thoughts.'

'Then I thank you. But he has sent medicines.'

Covington showed a physick. It was a thick glutinous mulch reeking of sulphur, and Phipps was humbled.

'I must take care to be charitable with him,' he said sincerely.

'I love you, John Phipps, you are the same as my own kin to me, even in our storms.'

'Now it is my watch,' said Phipps, bunching his beard in his fist on hearing the bells.

'John?'

'Yea?'

Covington reached out and took his friend's hand. 'Did you pray for me, John?'

'I did that, and I believe you were pulled through by a power so great that we are plain sinners to have bad words.'


The next to call was Volunteer Musters bringing his news. He said there was a great argument between Capt and Darwin while Covington thrashed in his hammock. It happened two nights before. At the end of it, Darwin declared, he would no longer dine in Capt's cabin, but would join the gunroom for his gut-pudding. The gunroom was made gloomy by the division, for their voyage drew its mood from the wedding of Capt with Darwin, and the gentry were like children when their parents scrapped, uncertain of their place. By the evening of the next day the two had patched it up again. So what was the fight about?

'You don't like to hear,' said the child Musters, looking at Covington with sly contempt.

'Why "don't I like to hear"?'

'Shan't ever tell you.'

Covington leaned over and twisted Musters's ear.

'Quick, or I'll make you cry.'

'And I'll have you flogged.'

'I am strong for that.'

'It was slavery.'

'Whose slavery?'

'Why, slaves' slavery. Whose do you think?'

'And you were on our captain's side,' said Covington, letting go of Musters's ear. 'You are all for slaves.'

'Positively so,' he said, rubbing it, 'otherwise I am a mutineer.' As a parting shot Musters said: 'And if I was a captain I would have you hanged, Nob Head.'

'And I would have you served for supper, King Dick.'

They sailed clear of Bahia letting go the lustrous
sights and stinking fevers of the place. Covington
was weak as water but up and about on the deck
and doing his scrubbing. Mr Wickham came by and told
him to cease and take his rest by order of the Capt. But
when Covington turned to seek Capt's eye he saw Mr
Darwin looking at him. So he tugged his forelock in that
direction.

They nudged south for Rio in sparkling seas before
gentle winds. And soon Covington's head was freshened,
his hopes grew fat, and he wasn't the only one in such a
mood, for the whole crew was singing as if not just
Covington but all his brother-tars were given the kissing
crust by order of the highest.

Wonders began in those Brazilian waters. Dust blew on
them over the sea from Africa. They came into a mirrored
bay. The sound of insects could be heard from the forest a
mile away. Creatures Covington had not observed in
boyhood voyaging now sprang plain to his eye. There was
a fish that inflated spherical. A cuttle-fish that glowed in
the dark. The whaleboat came alongside with stones in
sacks, taken from an island, and Covington saw men pass
curious shapes down a companionway. He felt a jealousy to
know the meaning of the cargo, but was too weak to go
ashore where measurements were made, specimens were

taken, and eggs were stolen from seabirds' nests. He was told by his shipmates that all hands were employed in making April fools, and felt like a fool himself. Indeed when midnight came they played true to their word, and nearly all the watch below was called up in their shirts, carpenters for leaks, quartermasters because a mast was sprung, midshipmen to reef topsails. Even Darwin was tricked by being asked if he had ever seen a grampus, a creature that didn't exist.

By listening, this is what Covington learned. That though they were afloat on the sea all was one with the land, and islands were mountains, and time stretched vaster than could be dreamed. Such knowledge ran through their bark and all seemed touched by this buzzing of news aboard that was never heard in the ship of Capt King. It was not like excitement, for there was no special time when it surged and faded, it was just a present thing, such as pride or joy. It came from the wedding of Capt and Gent, in their many disputes, hard-won agreements, and excited observations of wind, tide and stars.

Capt with his fine-made chronometers and weather instruments and loud voice postulated to the Derbyshire gent whether the world was made this way or that, and the midshipmen sang True or False, that rocks are a lid on the earth, let us have wagers, and when the Pot Boils the land spews Hot Coffee (their name for FitzRoy).

Darwin passed Covington once or twice, in his hurry about his work: 'I see you are better, Sailor Covington?'

The manner of the naturalist was full of knowledge and understanding. Say that now he had his eye on Covington, for sure. They talked about Bedford, the tradesmen who preached, and all manner of texts that Covington knew by heart, and how boys raised on Bunyan loved to split their meanings as smartly as the old Hebrews. Covington was not entirely well yet, but grew stronger for the asking after him. Although about his trowsers no more was said, and he would have to give the matter a nudge if he wanted satisfaction.

Covington joined Volunteer Musters for his turn at the bowsprit. Musters's hour was the middle watch when the rest of the world slept. They were among the stars. Chirpy as a cricket Musters rattled off bearings. Capt and Gent were above, wakeful and talkative. Between them it was all, 'I do declare!' and 'Nay, it can't be so!'

Covington shifted back to hear them better. Wee Musters became cranky with such eavesdropping and gave him his frown. Musters's family was neighbour to Darwin's uncle in Derbyshire, giving Musters special understanding of Darwin, naturally, indeed an exclusive contract to all information on the matter. The voices of those lords of stars and insects came and went. Covington sometimes drowsed. The ship surged and lifted. On a great beam reach it was as if they sailed through heaven, propelled by a creaking and shuddering down the length of timbers.

Musters made a sound of keening in his throat as boys do when pretending deafness, raising himself like a grasshopper on spindly legs. He considered it unseemly for one of Covington's low rank to witness dissension in his betters. 'Whatever,' said Covington, bidding the boy goodnight, making his way to his place above the coalhole where he rose and pitched, and rose again slung in his hammock, and wondered about himself unto sleep, considering that to rise was a requirement of the gospels, and that the rising overcame the dying.

*What of your fortune?* wrote Mrs Hewtson when they arrived in Rio under the Sugarloaf rock, and were given their mail. *Are you hard after your learning, my big man, Tom Noggin?* Covington took her letter to the rail and hung over the lapping waters committing the few pages to heart, remembering the crowded kitchen with its amber light, the clean scrubbed faces of the sprogs waiting for their Pa to come home and embrace them, after which he would turn and hail Covington-in-the-corner. Mrs Hewtson knew Covington's heart; he knew hers; and there was no place for him in that world of Mill Lane and the few streets nearby. He wanted to boast he was through to his next expression of life, but on turning the reverse page of her letter wondered if he was—for he found himself weeping, learning that his own letters were not arrived in England yet, and that even after all this time Mrs Hewtson believed Joey Middleton to be still alive: *Your Joey, how is he, you must be all so grown?*

What Covington recalled of Rio from boyhood days was a goat with a bell; a priest in scarlet robes they threw stones at; glorious Marys in gold leaf that made him ashamed to be a watery Congregationalist; a seawall under a thorny tree; and a gathering of poor naked black children to joy with 'Hosay'—with Covington striking his fiddle while Joey danced and spat, and rubbed at the cobblestones with his broken shoe and danced till his knees ached.

Covington went ashore where the cutter landed him, near the great cathedral. He wasted no time taking interest in his surroundings any more, nor in the rioting of his mates as they sang and boasted and fought their way forward onto land and through doorways and were gone. Nor in the louring looks of Phipps, who was in a mood to go catch them a cockfight in a place halfway round the bay. There was no pretence in Covington's isolation from that charmed company of saints and devils any more. He treasured his understanding of what it meant to be at liberty on the face of the planet. He was making an end to his life as a sailor, if he could.

In a tailor's shop Covington obtained his satisfaction according to a law fitting his station. It said that if a gent paid for the repair of your trowsers, then that same gent acknowledged your service and liked it well, and would call upon your service again. Covington paid for the repairs with his own coin, acquiring a scrap of paper marked with a cross as proof to the man who spoke of recompense.

Before Covington returned to his ship, however, he went up a lane and onto the brow of a hill, where a walled villa caught the breezes. He kept looking behind him, braving the presumption of what he did. It was an address written on a card in Mr Earle's fine hand—an introduction to a buyer and seller of birdskins and butterflies, animal bones and who knew what else besides. Covington clapped a bell and listened to the rustle of banana leaves overhead and heard footsteps coming to enquire his business.

'I am here to see a Mr Beskey,' he spoke through a hatch.

A question was asked of Covington in Portuguese.

'I shall tell you my *vendi* and my *comprar*,' he answered, 'if you let me in.'

A bit more of such banter and they let him through.

**B**ack on the *Beagle* at nightfall, Covington waylaid Darwin, telling him what he had done about those trowsers that had been promised him for his trouble before he went sick.

'You took this upon yourself?' The gent showed irritation and a grudging regard as Covington displayed his patches.

'I did,' Covington produced his receipt. 'I can be constructive,' he added, passing a hand over the bumps on his head that promised his fate, 'no matter what is said of me.'

Darwin allowed a smile. 'Well, I am getting to know you, sailor.'

Covington slightly bowed. 'It is just a mending job and very cheap.'

The son of the wealthiest man in Derbyshire promised to fetch the boy a Spanish dollar from his cabin, for which Covington might cover the repairs done, and—the gent clearing his throat, emphasising the cost—equip himself with a new pair of trowsers cut from black duck as well. Perhaps two? 'A dollar shall be plenty,' Covington spoke his gratitude, 'for three, if I choose my cloth careful.' And so they parted.

An hour later Covington was sent for and reached the poop cabin quick smart.

'Sailor Covington. I have your money for you.'

Covington caught the coin in the air and clenched it in his fist. Once more he spake his thanks, beholding this boyish red-faced and cautious benefactor surrounded by inks and papers. That was all, it seemed, and yet Covington was not quite ready to quit the cabin and nor was Darwin ready to let him go. Both stared. The gent shifted himself in his small chair and said:

'Well? Are we finished?'

And Covington said: 'I cannot say.'

Gent steepled his palms as if to say, 'Nor can I.'

Covington knew the importance of the next thing even before the gent said it, for he was a twitchy hound, and used his nose—preferment hanging thick in the air—whereupon Darwin cleared his throat and tried to make everything considered seem impromptu, which was hard for him, since his thoughts went ahead of him like the bow-wave of a rowboat in calm water and smacked Covington in the nostrils.

'Why are you grinning?'

''Cause I am ready.'

'Sailor Covington.'

'Yoi?'

'When the time comes, will you carry my guns?'

Covington thumped the chart table and wedged himself face to face with his benefactor—leaking spittle from the corners of his mouth. 'Upon my oath I will. Your guns and your bird-baskets and what else besides.'

'Not so fast,' frowned the gent—Covington lacking nothing to complete a picture of devotion. 'You are talking about too much work.'

'That is nothing to me.'

'I said just to carry my guns. And then only when the captain allows it.'

'But you have far greater need for a servant than guns.'

'There is much work that I have not fully determined as yet.'

'I shall help you decide.'

'Possibly.' The gent could barely disguise an edge of derision. 'What pray is your skill in natural history?'

'Nothin', except I can skin a rat and prob'ly a beetle besides. It is a rule that follows—the more work a servant does, the better work there is to be done.'

'You make it sound as if the tail wags the dog.'

'Nay, but I am loyal. Ask anybody going back to our two ships, the *South Sea Castle* and the *Adventure*. I am an old Patagonia hand, as they say.'

'You jumped in the water, I heard.'

'I did,' said Covington, somewhat startled. It was quite a time back. It was among the men.

'The story is told by Captain King.'

Covington felt himself colouring. 'Is it indeed,' he clenched his jaw. He had wanted to be known; he had better wear his colours with pride.

'King is my beau ideal of a captain,' said Darwin, answering Covington's stare.

'Well, as I say, I am ready,' said Covington. 'In taking to water or carrying your guns I am ready as anyone could be.'

'Learn this of me, Covington, I do everything in small parts.'

'Yet the whole ship sails for you,' said Covington.

'That is nonsense. She is on a survey.'

'I do remember that,' said Covington with a smugness that made Darwin frown.

'There is a matter of getting your release from the captain.'

'It will be easy. Capt hates my face.'

Darwin's eyebrows rose and he irritatedly ran his pink tongue across his lips. 'For goodness' sakes.' Then he cleared his throat, and climbed to his feet turning a deep angry red. 'You endanger your employment with a loose tongue.'

'It was only gossip I heard. Capt doubts that he loves me.'

'Must a captain *love*? We are not in a Bedford chapel, Covington, we are not down to splitting texts. We are in a ship of His Majesty's Service.'

'Pardon me, I forgot that we were,' said the boy, keeping the derision from *his* voice this time, and meanwhile being sure to etch the Bedford meeting house in the upper part of his mind where he etched the Celestial City.

Darwin sat down again and his face paled. He rapped his fingertips on the chart table.

'I am keeping you too long.'

'Yes, I have scrubbin' to do.'

'There is just this with me, Covington. Please remember it: I will begin to doubt my decisions if you niggle me.'

'You shall doubt me never,' swore Covington.

'That is all we need between us, then.'

'Indeed,' nodded Covington. But still he waited.

Darwin said, with a sigh, 'What is it?'

'You must tell me yea or nay,' said Covington. 'Am I to have the work?'

'*Must?*'

Darwin was affronted all over again. To his landsman's eye it was the romance of all sailors to make free, yet quite incomprehensible and hardly to be credited. 'I shall have to be sure of you,' he said, blushing as if in a courtship—and doubling around on all his unmade promises in an instant, it seemed to Covington, who stepped back in the doorway and radiated touchiness.

'But you *are* sure of me,' Covington returned. 'I have had my proof. You praised me in this very room—and see—I have my trowsers on.'

'I hear you have made an insect collection,' said the gent.

'Have I?' said Covington after a long pause.

'You boasted it to Mr Earle an hour past.'

'There is no harm in that,' Covington swallowed.

'How curious, though.'

'"For a sailor", or what?'

'Indeed, "for a sailor"—crossing the seas with beetles and grubs on the waves. Note I am out on the deck every day hauling butterflies and bees from under the ocean.'

'You are laughing at me.'

'Well, I am sorry.'

'There are grubs enough on our bark,' said Covington defensively. 'Weevils and mites and cockroaches. I can make money from the art if I try. That is what I am told. And anyhow,' he slyly added, 'you have been aboard when a butterfly came through. I remember you chased it along with your net, and then Midshipman King sat on it, didn't he?'

'That is true.'

'And that collection, it was not made by me,' said Covington stoutly. 'I went to a man and bought it. A Mr Beskey, a naturalist. Our artist told me where to go.'

'Full marks to him.'

'The frame was all broken, so it was cheap. I fitted it together again in a jiffy, and now I have one, and more to the point as they say—I know how it is done.'

'Leave me now,' said Darwin, and when he spoke those words there was no disobeying him at all.

Covington dipped his head, said, 'Very well,' and drew his shirt around him, stepping backwards through the door into the dark.

# BOOK

## 3

### *On Some Useless Afghans*

# 1860

There seemed no end to the confusion Dr MacCracken felt in those early months when Mr Covington bought the cottage 'Coral Sands' and came to stay, establishing himself, with Mrs Covington, as a fixture in his daily life. If ever a man was made to stake out and invade another human being's territory it was the brutalised old bruiser, as MacCracken dubbed Covington, with his whiff of mystery and resentment in about equal doses, and his deaf-man's blundering stare.

The month was April, Sydney's autumn, with the first cool nights wafting up from the south and pockets of heat through the day making for a glassy stillness on the waters of the harbour. On starry nights the only civilised companions for a man were a bottle of claret, a portion of ripe cheese, a volume of Keats and a proper sense of melancholic withdrawal from outer distractions.

But it was too much like an island where they lived, and MacCracken itched at the possibility of running into Covington everywhere he went, of having him come to his house at whatever hour pleased him. Yet what could he say? MacCracken's temper must take a lesson from his dog, and 'do when told' because of the dependence that had developed between them.

The profitable business MacCracken ran with Covington meant he could not afford to be rude, even when Mrs

Covington acquired for herself a milking cow, and when her dairymaid—and sometimes Mrs Covington herself—led it clattering past his front porch, where it swiped his mint bushes and deposited cowpats on his Italiano tiles.

Beginning by feeling a frank affection for the man, MacCracken came to feel he might strangle Mr Covington pretty soon. He began to wonder if Covington's harbourside self, as it now somewhat daftly declared itself—canvas trousers, linen shirt, bare feet, floppy hat woven from cabbage-tree leaves—was a sign for MacCracken to give up his place and shift his horizons. The man was too close. The man was a confounded nuisance. Men like Covington had no easy retirement about them. Yet there he was, an avuncular Napoleon in his exile. And for why?

After playing skittles in his head on the question MacCracken decided it was up to Covington to make a change and ease his irritation: *he* wasn't going anywhere. His pleasure in Watson's Bay was blunted, not sated. His days were not demanding. He loved his daily swims, which he took with a boy named Charley Pickastick to watch for sharks. He dreamed into the night over many a page from his well-stocked shelves. And he tarried with Miss X, who pleaded little of him except to tweak his ears and twist his hair in her fingers and ask, 'Am I not beautiful? Do you not love my eyes? What colour are they, Davy?'—to which MacCracken readily gave proper answers, thus gaining her most intimate favours. Though when she asked him if he loved her he became cranky, and criticised her on some small pretext, such as the hour she arrived or left being a nuisance to him. He reminded her that their arrangement was gratifying to them both—sworn in delightful secrecy at a midnight rendezvous. And so what else did she want from him, that made her demanding of a sudden, and unattractive with tears?

'I don't know,' she pouted.

'Come here then.'

She nuzzled close.

Then he smiled to himself, being quick at putting her at ease, pleading for his good opinion, and getting all back as before—close and generous to his desires.

MacCracken would say just this, however, if pressed to expose a hidden-away part of his honesty: that his life had an over-calm suspension about it, being neither passionate nor fraught. It was at his father's insistence he had studied medicine and through coming near the bottom of his class had found himself practising surgery. Though he disdained the sawbones' trade, a rote mastery of anatomy and the sure hand of youth meant he kept busy, content in the knowledge that a few days' hacking on broken frames each week secured his name under heaven, so to speak. But Covington? There he was tapping on MacCracken's window pane at an ungodly hour.

'Who is it?' reacted Miss X in alarm.

'It's Mr Covington,' MacCracken answered her. 'Stay where you are.'

'What does he want?'

MacCracken was ready to wound him. 'Don't know.'

He took a candle to the window and signalled Covington away. *Go, get off with you. Yes now immediately.* Covington had the insomniac's look about him, the frantic obliviousness to time of those who live in their thoughts. MacCracken opened the latch and through a gap in his door of a mere inch, that widened, Covington began speaking in a rush.

'Has there been a package addressed to me?' he demanded to know, 'that might have come with your mail by mistake, MacCracken, do you remember it at all?'

The doctor shook his head.

'Has anyone said, "I have something that isn't mine, a book that isn't mine, a strange book, perhaps, a frightening book that a decent person might wish to hide away"?'

MacCracken shook his head again and rolled his eyes,

making a sound like a mournful owl in mockery of his stone-deaf caller. 'Good *night*, Mr Covington!' he shouted.

'Nobody speaking of the devil's work,' Covington persisted with his line, 'more than is warranted? And mentioning my name in that connection?'

MacCracken raised his eyebrows. ' "Covington"?!'

'Certainly. "Covington". A name that might be written in that book. Doubtless in small print but written there all the same, MacCracken, as an acknowledgement of, ah, existence. Damn me for an imbecile, man, and just tell me. Any book?'

'No!' MacCracken yelled into Covington's eyeballs.

Covington's boot stayed thrust in the door and his glance went darting zigzag towards MacCracken's library where an inner door stood ajar, and Miss X's naked toes showed pink and glowing in the light of the hearth. MacCracken saw that Covington on sighting those feet forgot his book altogether, and took to the dark somewhat enlightened.

Down on the shore one day, at the lowest ebb of the tide, MacCracken saw Covington kneeling with a short blunt knife and flicking barnacles into a basket. His motion appeared leisured and unhurried, yet his basket soon filled. MacCracken forbore from malice to say a word about Covington's choice of dish, which his noble blacks, being connoisseurs in such matters, disdained as unsuited even to nature's table. Nonetheless Covington filled that basket. And the next. So MacCracken mimed a coin between thumb and forefinger. *What will you get for those?*

Covington looked around to see if there were ladies present, and seeing there were not, made an obscene sign of disgust. 'I'll get less than nothin',' he grunted.

'You're right about that, Covington old mullet,' MacCracken said. Malice whipped him along and he scored an edge to his pleasure as he squatted on his heels, because he knew Covington couldn't hear him. Yet to his absolute surprise as Covington sorted through his load he responded, 'Fish?'

MacCracken stopped cold and looked quizzical. Who had mentioned fish?

MacCracken himself had.

'You called me a mullet,' said Covington.

'You *heard* me?'

Covington turned his back and acted as if nothing had been said.

The day continued as it had begun. But MacCracken was stupefied, and retreated in thought, leaving Covington to the shore. So Covington could read lips. So he'd kept the secret from MacCracken this long time past—an ace up the sleeve in his dealings with men, and doubtless the key to his fortune and much else about him besides. Here was a man of secrets wrapped in a secret indeed.

That night MacCracken took hold of himself. So that when Miss X (let him not be so coy—Georgina) tapped the glass, entered his study, and found him under a pool of lamplight, she saw him engrossed in his native philosopher, Ralph Waldo Emerson, within whose psychology he retreated as others went to their prayers. He hushed her, making a 'get thee hither' gesture. She knelt at his knee. 'David?' she plucked at his trousers. 'Appleblossom?'

'*Please.*'

'Am I still your special annoyance?'

'Yes. Indeed. Of course. Whatever.'

He needed to be removed from his own vanity. Emerson was his guide for the moment. His enmity towards Covington subsided under the balm of a wider wisdom. And what on earth was that irritation based upon? MacCracken's own nature most certainly. He was enough of a student of the mind to know it, and yet enough of a human being to be caught by the feeling as stickily as a fly gets caught in a web.

A few days later MacCracken went walking the cliffs on the ocean-side of the Heads and saw, away below him, treading the eyelid of the great Pacific, the strange figure of Mr

Covington balanced on a dripping block of stone. He was at his barnacling again. But at what risk!

The sea in that place was a wonder of the world. It rose in a bejewelled surge containing many tons of water, and sank rhythmically, relentlessly, draining enormous square boulders. All was tide. All was deeps. All was the hidden mystery upon which MacCracken's subject took his toes— protected by canvas shoes—and dared his unfathomable life.

Note there was no land between this point and the desert coasts of Chile four thousand nautical miles distant. Between falls of water—white, green, all streaming in gut-tered torrents in a heaving Niagara—MacCracken saw Covington stepping out and back, prising shellfish from rocks in his studied manner. From directly above, MacCracken's eyes followed where Covington stowed his baskets on a ledge. He came to them. He reached up. He spun around. He flattened himself against the honey-brown walls as a wave came and wetted him hard. Then out he went again, reaching with his long arm into crevices rumbling with power. MacCracken saw him remove a spiny sea-egg, hold it to the harsh and twinkling light, and drop it back in the water again. The place was so dangerous that nobody save wrecked sailors ever stood there, to MacCracken's knowledge—and then in great fear of their lives. He was ignorant of Covington's purpose, having dismissed the possibility of rare and valuable collections being made for sale. There was no beauty in what he scraped. Strange that MacCracken gave no thought to Covington's courage in his lonely chasm. Nor that he might have intelligence as his guide. Nor that the book that haunted him was real enough, representing a danger to Covington greater than any cliffs and raging seas—the nails in his coffin, so to say.

How the fellow had descended the cliff MacCracken did not know; and now he waited to see the route of his climb

back up, because it made him laugh, and thrill with anticipation to think of his managing it at all. But even
witnessing the climb he was unable to tell the way
Covington did it. Hand over hand, balancing like a goat,
dancing like a monkey, he confounded the picture given so
far of a stolid figure of scarred flesh and crippled blood.
One moment MacCracken hoped he would fall, and cease
his niggling ways. The next he slapped that devil in his
mind, and spat in its face: and there Covington arrived festooned with baggage and limping the clifftop path,
extending a slimy hand towards him, 'MacCracken!'—the
other hand running with small cuts as Covington dropped
his baskets and wiped his forehead leaving smears of blood.

'What a show you gave me!' mimed MacCracken, and
then, mindful of Covington reading lips, repeated the same
sentiment aloud, with a more careful shaping of his mouth
for Covington to read as skilfully as he chose.

About Covington was the air of challenge that prickled
MacCracken so. His voice boomed like the sea. 'I don't
ache when I'm working,' he confessed, giving his equipment a kick. 'That's one good thing about 'im.'

He began unpacking himself onto a flat rock baking in
the sun. MacCracken wanted to ask, 'Him?'—but before he
could speak to Covington's face—wanting to try his lip
upon that eye—he was barked an explanation:

'The man I send things to. From time to time. If it pleases
me.'

'Of course, "If it pleases you", so on and etcetera,'
MacCracken prattled behind his back.

From around Covington's knotty shoulders was slid a
canvas knapsack. It was furnished with four tin boxes like
anglers' worm boxes made of different sizes. Into them
Covington sorted various shells from his baskets: small
black and shiny ones, others whorled like the spirals of a
straw hat. The best he retained were dullards. Yet he had

risked his life for them. MacCracken leaned forward, keenly following his choice.

Covington wore a kind of quiver made of wood resembling a butcher's knife-holder. It was strapped to his waist and contained a small hammer and chisel, plus an assortment of oyster knives. As he squatted and packed the shells, leaving the live creature within, completing his 'work', as he called it, MacCracken saw him as a salt-crusted smithy or indeed a butcher of the natural world. It was MacCracken's first intuition of the simple truth and a waypoint in the journey he was already embarked upon with Covington that would change their two lives absolutely.

'That man I send 'em to,' Covington grunted. 'I call him La Naturalista. You know your Spanish, MacCracken?'

'¿Es posible cruzar estas dos especies?' MacCracken quipped, pointing to two different shells, aware that Covington's eyes were very much upon him.

Covington gave a dry laugh. '"Can you cross the two species?" Aye. You're just like him.'

'Your naturalista?'

'Some might say so. A man who knows everything. That's you,' he faintly curled the end of his smile ('I daresay in honest admiration,' thought MacCracken).

'But of whom do we speak?' MacCracken teased at him. 'A Spaniard?'

'What's that you say?'

Covington looked MacCracken in the eye.

'A Spaniard. Your Don Sia Di,' MacCracken enunciated.

Covington turned aside. 'A Spaniard in treachery, may be.'

A familiar bitterness overcame Covington. MacCracken had observed that curl of the mouth often enough to assay its quality. It was not loathing or malice but more a response to the aftertaste of a medicine.

'Covington!' he tapped him on the shoulder.

'Yoi?' Covington turned and faced him again.

'I know you understand what I say,' he said, enunciating just a little more slowly and carefully than usual.

'That I do,' said Covington, with such simplicity it sprang tears to MacCracken's eyes. 'You swear oaths behind my back, and belittle me with cheap gibes.'

'I shan't betray your confidence on this matter,' MacCracken swallowed, ashamed.

Covington shrugged as if to say, 'Well, if you did, it would be your conception of honour, not mine. And hang you for it.'

Covington held an assortment of shells out to MacCracken. 'See the differences between them? Wouldn't you think, in their eternal variety, that God had printed these creatures, each and every one?'

MacCracken had never looked closely at the common limpets before. To do so humbled him extremely. He saw that each had striations and lines as different from each other as the finger-pads of human beings. In their very dullness they were profound.

'So if you were looking for bounty of God,' continued Covington, 'where would you look?'

'Into these very shells,' answered MacCracken, putting himself in the role of pupil to Covington's hoary old philosopher.

'Well, that is why I send them to my gent.'

'So he can find God?'

'You could say that.'

'Is he mad?'

'It may be. But I fear not.'

'Are *you* mad?' was the next, unspoken question on MacCracken's lips.

They walked along through the heath together. 'My Afghans are coming on Friday,' MacCracken said.

Covington wheezed with amusement. 'Your "Afghans",' he smirked, and in that smile MacCracken began to like him again in the way of old friendship, where much is shared in common and easy dismissals are not so ready to hand. It was the same grin of delight MacCracken saw when Covington sighted Miss Georgina's bare toes.

'It is my turn to have the Afghans,' MacCracken said, 'and I want to do them proud.'

Those Afghans. They were a quartet of good-humoured dining companions. Whenever they visited, MacCracken arranged for them to stay the night, retaining Mrs Franks, his Belgian housekeeper, to prepare them a great dinner, and organising a fishing-yawl for the next day. They had this in common: they had met on the Ballarat goldfields, and failing dismally at gold-getting had walked four hundred miles back to Sydney through a bushfire summer. On their arrival, smutted and bewhiskered, they plunged themselves into the cooling waters of the harbour, swearing never again to stray from metropolitan delights—nor from each other's firm devotion. At their dinners they wore a variety of fantastical hats. Covington had seen them sitting on MacCracken's verandah one evening, O'Connor (architect) as a Turk, Evans (bookseller) as a Scot, Forsyth (schoolmaster) as a Sikh, De Sousa (shipping agent) as a Malay—and MacCracken (sawbones) as a courtier in the days of Versailles.

'Ah, yes,' Covington chortled. 'What *of* your Afghans, MacCracken?'

'Friday is the day. Can you get us some of your shellfish —fresh? You know, I rather fancy scallops,' MacCracken said, terribly aware, at that moment, of an appalling breach of etiquette with Covington, whose face turned more purple than plums.

'Scallops?'

'Sure.'

'Scallops, are they the ones with twelve little eyes lookin' at you?' he sneered.

'I believe they are,' MacCracken lamely responded.

'No, I will not get your scallops. That is not my meaning, Dr MacCracken.'

'I see. Thank you.'

'Just as *your* meaning,' Covington thundered on, 'is not to take a man's leg or arm from off of his body, or remove his appendix, even, for the mere benefit of his asking.'

So saying, Covington stood for a few moments in the bright sun of hot Australia, salt spray drying on his face like scaly skin. He put MacCracken in mind of a big clumsy insect that had just clambered from its chrysalis.

'I am seeing the man for the first time and have misread him absolutely the whole way along,' MacCracken thought. 'He is not set or stubborn but twisted out of himself constantly.'

Without further words Covington re-packed his gear and they followed their separate pathways home. MacCracken burned with shame as he watched the man pick his way through the bushes, and then, dropping below the wind-scorched plateau, enter a sheltered gully where he slipped from sight, neither raising an arm nor waving his hat to MacCracken as he usually did. MacCracken had treated him as a servant and got his reply.


So, MacCracken wondered, how to begin putting things to rights with Mr Covington? What *was* his meaning at all?

It seemed that Covington wanted to tease MacCracken into an understanding of him bit by bit. So they were bound somewhere together. Mrs Dorothea Covington had said as much: *I do my best, but he has such hopes for you, Dr MacCracken.* Yet the man's revelations, such as they were, resembled slabs of stone exploding from an ancient monument and crashing around his feet. There were long dusty silences at intervals. Meantime apprehension swam around his head like gnats in the evening. He was always demanding of MacCracken that 'something more' with his imploring, fate-smitten eyes—something he asked MacCracken to grant him that was beyond MacCracken's powers as a doctor, or as a man, to furnish.

Which was what? Release from something articulated inside him but unspeakable to the outer world? What a barnacle Covington was in himself. MacCracken felt only the outlines of Covington's soul when he pressed with his mental instincts, feeling how it clung, sucked, wheezed, dripped, tightened. But also shone. That had to be said. When the tide fell, and life was left clinging, in the sunlight it shone—with persistence, sadness, puzzlement, ire. They were qualities worth gold in the man and not the gold of his commercial dealings either.

What on earth was he to do, having done the physical part with his scalpel, making Covington whole in flesh from his brush with peritonitis? That, it seemed, had been the easy road. Now for the mental part: it would be harder. In relation to the soul—if it was the eternal soul that Covington's anguish meant—MacCracken had come to a full stop in the dissecting room. He was the wrong man for the subject. He could give no assurances of eternal life—no, never—and though he was a declared Episcopalian the idea frankly appalled him of a bunch of angels in grubby night-shirts mouldering till the end of time with no prospect of change. It didn't fit any picture of life MacCracken had. If life was eternal, wouldn't it have to have *life* in it? And where was life, he asked himself, except in blood and breath?

Yet it remained that he was chosen and it remained that he felt himself obliged, and it remained that he loved the old stager too. So MacCracken decided to make himself into a better man to do this thing that his friend demanded of him—emerge from *his* egg or his chrysalis or whatever it was that gave him his protection at the age of twenty-seven for pleasure-mongering in extremis and hollering 'hang the rest'. It was time for MacCracken to seize his vocation of the mind and start hauling it in tight. 'Old horse Mr Covington, old gristle-factory and barrel of pride,' thought MacCracken. 'You are going away from us all. And how do I know? *Ask me if I know when the sun goes down.*'

That was the instinct MacCracken had now. And if he could he would follow Covington down.

To Friday's dinner—where Afghans ate like beasts and quaffed wines like water at life's eternal springs. Whenever Mrs Franks entered the room they minded their manners and when she left they shed them again. She did them proud with her table setting. The smells coming from her kitchen made them groan.

Once they had done a perish. As failed diggers the bunch of them had eaten scraps and roadside findings. They had tempted themselves with boiled thistles, pumpkin tops, pigweed-lettuce and the like on their march to Sydney and now they demanded their dues. They had eaten wombat, possum, snake and wild duck so devoid of flesh they only sucked bones and were glad of it. Now they had candles on the table, crystal goblets, pewter bowls. They supped on calves' feet broth. They had beef braised in cider, with a grand name in French, and hot pastry delights stuffed with silver trevally and small juicy prawns. They had salad balls with honey dressing, and this is what those delicacies were, said Mrs Franks: a concoction of butter, lemon juice, cayenne, curry powder, and cheese from Mrs Covington's dairymaid, all rolled in chopped parsley and served on a bed of green lettuce. For their sweets they had hot sponge with lemon sauce and French jellies.

Finally after Mrs Franks cleared the table for the last time, balancing plates in the dimpled crook of her arm and chirping about boys who believed they were men, they chose their headgear from a box of hats, and MacCracken passed around the port.

De Sousa wore a black fez with red spangles. He asked about Covington. He'd spotted him during the week dining with officers at the barracks:

'Quite a pack of fawning marines and naval men. I believe your Covington lends them money and affects up-country frugality—mends his own boots, and so on—but is all sham and pernickety side. He's rich as a Rothschild and his one motive in life overriding all is to get his daughters husbands.'

'Could be,' MacCracken murmured, giving nothing away, for he had nothing much to give when it came to talk of daughters. Covington rarely spoke of them. MacCracken tugged the strap of the overlarge policeman's helmet he wore, and announced a list of fines to be charged for various misdemeanours at table.

'"Indecent speculation"—sixpence. "The betrayal of eternal friendship"—one shilling.'

'Daughters, where *are* they?' O'Connor interrupted. He twirled his 1844 Penfold's muscat to the lamplight, and spoke from under the shade of a coolie's hat, wide as a wheel, that was whisked from a Chinaman's head at a roadside camp and favoured by all of them ever since.

'Ah, the daughters,' said De Sousa. 'Fearsome ugly, I'd say. No-one has seen them.'

'MacCracken? Come on with you—what gives on Caliban's isle?'

Forsyth, wearing a shako with the Russian insignia of rampant eagles rising from the crown, interpolated vaguely: 'Covington? I don't think I know him. Is he the duffer I saw at your wharf, MacCracken, with busted veins in his hams and a tooth missing, and a pugilist's nose? He tried to sell me a cod!'

'He goes around,' MacCracken smiled, 'a bit sarcastically.' The selling of fish indeed. He felt stung. Why not scallops? MacCracken wanted to be thought wiser on Covington than he was in fact, an interesting vanity considering his whim to have him fall from a cliff just the other day.

A lull fell over the table. The matter of daughters was endlessly compelling. The Afghans were sunk in thought.

'Can't be ripe 'uns,' said O'Connor. 'The MacCracken would be the first to pluck fruit if they were.'

'He is lickin' his lips, though,' said Forsyth, 'and thinkin' lascivious thoughts. Fine y'self sixpence, MacCracken.'

MacCracken pelted Forsyth with almonds and O'Connor pulled his chair from under him, leaving him collapsed on the floor.

MacCracken knew that Covington's daughters, underage children both of them, were in the mountains beyond Twofold Bay, where Covington's four sons, Syms junior, Charlie, Eddie, and Alf (who was barely ten years old!) rode wild horses, hunted kangaroos, and trimmed the Covington estates of their beef and timber. It seemed such a conventional colonial picture, and as the Afghan rule was to be 'interesting', MacCracken munched on a dried fig and said nothing.

'Look at old "Mac-a-cracker", he knows more than he's sayin'.'

'Roll over Miss X?' murmured Evans. He wore a Mongolian fur with ear flaps, though the night was balmy.

'Back off,' MacCracken snarled.

'Sorry, dear chap,' said Evans. It was an awkward hiatus. MacCracken feared he offended Evans but dammit he offended himself. Out of five bold Afghans MacCracken was the one with a mistress, or dolly, as the word was with them. There was too little reality about MacCracken sometimes. He knew that Lizzie was twelve and Emaline was six, that the older made a fierce protective mother to the younger, and that when the time had come for Mrs

Covington to bring them to Sydney they had outright refused, and gone hiding in the bush.

'Daughter, singular,' said another. 'Her name is Theodora. Covington's been taking her around, walking the gardens. *I* haven't seen her, but they say she's a catch, red hair, high cheekbones, emerald eyes, and a graceful walk—a dancer?'

'Theodora?' said MacCracken, trying out the sound of the name and keeping his emphasis neutral. He knew nothing about any Theodora.

The eyes of the Afghans switched back to MacCracken. Having so recently penetrated Covington's defence against deafness he was now learning of another matter he'd kept from him all this time.

'You are all scallops,' he said, confronting so many beady stares. 'Shut your shells.'


The conversation passed to other topics, all of them muddled, hilarious, pugnacious and sentimental by turns, as was commonplace with their tribe. Later Evans attempted healing their tiff with gossip, and mentioned that his bookroom had a standing order for 'anything new in natural history' not just from MacCracken, but also from Mr Covington. A new volume had just arrived, *The Origin of Species* by Charles Darwin. 'I have it in my bag,' said Evans.

'Then give it to me,' MacCracken said.

'Now that's a bit awkward. I brought it over for Mr Covington.'

'You brought it over for Mr Covington?' carped Mac-Cracken. 'So much for friendship. You are fined a shilling, Furry Ears.'

'Look,' said Evans, 'I must be truthful. Your Mr Covington specially asked for the book by Darwin, "the one about everything" he called it, whereas your order was vague as mud. Therefore Covington has priority, I'm afraid.'

'Then give me another of it.'

'The whole stock of the title was spoken for before the ship unloaded. The last went to Krefft, the new man at the museum. He calls himself a Darwinist, whatever that is. Something German, I suppose.'

Darwin's book came up through MacCracken's thoughts like an object released from the seabed and defining itself ever-faster as it wallowed to the surface. It would answer why Covington was so insistent the night he tapped on the window, when MacCracken sent him away. In the style of their dealings MacCracken wanted an advantage over him in the matter.

'Let me have it, and I will pass it on to him.'

Evans went to get the book but changed his mind. It was too neatly wrapped in a rare, expensive paper and tied with knotted string. Damn him—why should MacCracken get his way in every department of life as he always seemed to? Evans yawned drunkenly and fell asleep in his room.

It was late, with the deepness of many stars overhead, when the rest of the Afghans staggered away to their rooms, swearing to be up early and fishing. De Sousa was the last to go and he plucked MacCracken's elbow. 'A final cigar?' The two of them sat on the verandah steps. The winking of native fires could be seen across the harbour. Water lapped the rocks at the foot of MacCracken's garden, and he heard—constantly against his back like a giant snoring— the sustained roar of the ocean.

A light burned late at 'Coral Sands'. With the clarity of imagination intensified by inner preoccupation MacCracken knew for an absolute certainty that Mr Covington was awake and touching the walls, listening to the amplified world vibrating in his bones. Indeed he had his proof of it the next day, when he found himself absorbed into Covington's dome of echoes, becoming part of it in much the same way that food becomes part of a body by being digested, and we are able to live.

Meantime, when MacCracken turned to his companion, he found that De Sousa had taken himself off to bed, and the cigar that MacCracken had thought was only just begun was a smoking stub in the grass. Time seemed to be moving erratically, like a firefly—here bright and then gone, and then over there again flaring up, with MacCracken lost in the intervals.


In the middle of the night MacCracken rolled over in his bed and came half-awake. He had his boyhood favourite, Charles Darwin's *Voyage Round the World of HMS Beagle*, on his shelves. Covington had seen it there and said that he 'owned' a copy. Then he had quoted from it by rote, making a special point over the word 'obtained'. With MacCracken the *Voyage* had played its part along with Melville's *White-Jacket* in inspiring him to take to a life of foreign travel, to bid his unbending father a fond hypocritical farewell, and so on and so forth. Thus when he left Boston he shipped down the coast of South America as Darwin had, tasting Patagonian gales in his teeth like so, and stopped at the Galapagos, and sailed the Pacific making his shell collection on the way. Looking back it seemed that Darwin was the reason and not gold-fever or character failure as he mostly thought.

MacCracken sat up in bed, a sorry grin on his face. He had Mr Covington at last. His *Don Sia Di*.

He sank back on his pillow, his head thumping somewhat from wine and his tongue tasting sour.

How could he have been such a fool? Through all this time MacCracken had thought of Covington's Don as a Spaniard, though Covington had never exactly said so, had only stated, plainly and in his own understanding of the matter, the identity of his patron—who was *Spanish in his treachery*—and linked him to the book containing revelations that frightened him.

So if there was something about Covington unseen, thought MacCracken, sinking back into his pillows, it was not through Covington's hiding it away. It was through dullness in MacCracken's instincts and his lording it over a deaf man.

MacCracken's own secret, that he was a dunderhead and slow-witted in taking time to twig to anything much, he would hide from the world until such a time as he set his own narrative down.

Next morning, before any of them at 'Villa Rosa' were awake, a boy drowned in the cove. MacCracken knew him as Charley Pickastick, his faithful shark-watcher. Down at the rickety wharf, whenever a boat arrived, Charley dived for coins. His cheeky face challenged visitors, his tongue making a high-pitched, excited wailing—a song chilling the blood when heard at night, and even by day not so much asking for reward as demanding propitiation—except that whenever Charley saw the affrighted faces of his listeners he gave a laugh, and rolled around clutching his sides as if Europeans were the greatest show on earth, and not wide-eyed simple black-faced he.

It was said that when Charley leapt off the end of the jetty that dawn his legs were crooked in midair and his elbows jerked no different from any other time. He was seen to arrow under in the shallow tide, and then spike the bottom and stay there. He was like one of the coins he dived for, glimmering fainter and fainter until gone and lost among waving waterweed where dugongs browsed and sea-horses dangled.

Roused by cries for help, bleary from sleep and with a head murky from port, MacCracken made his way to the jetty. 'What gives?' Fishermen were there, and a few blacks already grieving. 'Boy under,' he was told. Someone broke the water, gulped a great intake of air, and went under again. It was Mr Covington. He'd learned of the matter before anyone—who knew by what sign language and reading of lips?—and had run to the cove flinging his coat away as he went.

The Afghans joined MacCracken one by one, looking pale, useless, ridiculous in daylight.

'How long's he been down?'

'Ten minutes,' estimated a fisherman. 'Or longer.'

'Tell Covington there's no chance.'

He was told. He was deaf. He kept diving.

'The madman is an ox.'

Nobody else went in. On one of his dives Covington stayed down long enough to cause alarm on his own account. Then as Mrs Covington and others arrived he emerged treading water with the boy limp in his arms. He bellowed in the doctor's direction, 'MacCracken!'—surging from the sea bearing his load all strung with weed and disgorging water from the mouth. The crowd stepped back, forming a fearful, interested circle. Charley Pickastick was cold stone dead already as his neck was broken. 'Uh! There!' said Covington, plainly expecting miracles, having once been saved by MacCracken and expecting no worse remedy than the resurrection of a foolish boy. MacCracken made show of doing what he could, which was nothing at all except to be respectful in arranging the boy's limbs on the beach, and then to kneel at his side feeling helpless and to blame. Who was it had first flicked coins in the air and started Charley diving for them? MacCracken.

The Pickastick clan grabbed him from MacCracken's charge, running away with him rocking in their arms, giving out their wails.

'I've been waiting for this,' muttered Covington, staring hard at his doctor friend.

'Waiting for what?' blinked MacCracken.

Without ceremony Covington flicked his right hand as if

shedding water droplets. And then, bunching a fist, he slammed MacCracken in the jaw, knocking him over. Everyone heard the almighty slap—bone on bone—and MacCracken knew his jaw was dislocated at the very least. His head jarred, his vision doubled, and tears shed for Charley Pickastick splashed down his cheeks and stung. Then came worse. MacCracken twisted on the way down and knew his ankle was gone. Engulfed in a rush of pain he was plucked at by Covington, who grunted, 'That's better!' and MacCracken blacked out.

He woke to find the Afghans carting him up to the house and placing him in the shade of the verandah, with Covington sullen and trudging along beside him with his shirt draped over his shoulders. Cushioned in shock, half-dazed, MacCracken looked around at a crowd of helpless, nonplussed faces. Nurse Parkington was sent for and took command. At the edge of MacCracken's vision he saw O'Connor and Forsyth taking Covington by the shoulders, shaking him in an effort to get his teeth rattling. 'Idiot, madman,' they kept repeating, 'imbecile, why?' But they might as well have been dealing with a boulder for all the response they won. Mrs Covington used a sleeve of his shirt to wipe her husband's mouth, and admonished: 'You are such a one, dear.'

*Let him go*, MacCracken said to the Afghans, or tried to say, but found, as he spoke, that no words came out. He was without the use of his jaw. It was clamped tight as a barnacle.

'Take him to my house,' commanded Covington.

The Afghans stood in his way.

'What a capital idea,' said Evans. 'Quite perfect.'

'A very good one,' said De Sousa, taking a step backwards as Covington snarled at him. 'Excellent, I must say. What will you do there, bludgeon him a little?'

The Afghans went into a huddle. They were worse than useless if they wanted to save their friend. 'Might we send for Doctor Crews?'

They read MacCracken's eyes at this and understood: *Not unless you want me amputated.*

'Look here, why did you hit him?' They formed a circle around Covington but kept their distance.

MacCracken read Covington's eyes: *You wouldn't know.*

'We don't know.'

*Get back to your parlours.*

'Tell us why.'

Covington raised his fists making a show of them as products of nature: two rough blood-smirched stumps of a tree. *These'll tell you why.*

MacCracken, seeing those fists blocking the sun, was obscurely relieved that the knuckles were barked, that his jaw had made a mark on them. He hardly knew why, except to recognise that a point in his life had arrived when he was tired of who he was, and had come to feel a flimsiness of character running right through to the physical. Good to know he was solid, then, to some slight degree. He saw Covington's point somehow—how dare a boy's life be wasted when MacCracken was so lightsome himself, and fit for hanging?

Mrs Covington fussed between Covington and the Afghans.

'Do you mean to go with him, MacCracken?'

'Are you *sane*, MacCracken?'

They stared into MacCracken's face. 'Answer us, man!'

'He can't speak,' said Covington. 'On account of he's been hit "right-tenpenny" on the cranium. And as for any of you wanting to know why I struck him, he knows why I struck him, don't you MacCracken?'

MacCracken was mute. *Aye*, he indicated with a nod. He knew in a sour kind of way. He was struck because the boy died. Because nothing could be done about the boy. Because all the time MacCracken had known Covington the old dodger had appealed to MacCracken to know him—and MacCracken had not known him, refused, and saving the boy had much to do with it. That was the reason he was struck. He would never be able to explain it deeper than that—but he understood it. A boy had died. A boy had died.

Pain throbbed with the repetition.

'Well, old pal, are we to give you up?'

Adopting the logical imperative of his attacker that was in the process of carrying the day, MacCracken nodded. Nurse Parkington endorsed the decision. She was in cahoots with Mrs Covington anyway—they had become great friends—and she had often thought that a good hard knock was what MacCracken needed, and had sometimes rolled up her commendable sleeves threatening to deliver one.

'Stand back,' she scolded the Afghans. 'Give the doctor his air.'


Thus it was that MacCracken found himself surrendered without any great opposition to the care of the Covington family. Once the decision was made Covington himself was nowhere to be seen. It emerged he was over at the blacks' camp asking after the drowned boy, being the only one of them to go there and still in his sodden clothes too.

Before they set off for 'Coral Sands' De Sousa brought MacCracken his medical chest and Nurse Parkington fed him tincture of opium. When MacCracken was blissed he chuckled to himself. It was a nice thought he had. Covington was deaf and he was dumb, and so they were down to bedrock in their dealings with each other.

The Afghans lifted him onto the same planks that had carried Covington to him in the first place. As their last duty in their friendship they bounced MacCracken to

'Coral Sands'. Since the night of the rat on the table MacCracken had not been through that door. There had been invitations, but when MacCracken hesitated on hearing them they were swiftly withdrawn. Lamps had burned in the separate windows flickering offence. But now MacCracken was peeled of any pretension and so he was taken in. As he entered the house he rolled his eyes around. It was even more like a ship inside than previously—everything neat, scrubbed, fresh and somewhat threadbare, yet purposeful withal. He was taken through to the room matching his library (both cottages being of similar design). He found that Covington's room was a library too, reflecting MacCracken's own down to the many books, the chronometer on the wall, and the chart table in the corner—but also including the most beautiful displays of butterflies and beetles MacCracken had ever seen. Lying in a row on a tray were six rosella parrots, plump, stuffed, lifeless, each on its back and with its stalky feet in the air and tied with twine. In the opposite corner, in the matching alcove where MacCracken kept his globe of the world to spin with a finger and conjecture where he might alight next in his flutter of existence, Covington had a tin hip bath and a scrubbing brush and a bar of yellow soap.

MacCracken was placed on the chart table, farewelled by his friends' babble, and Mrs Covington and Nurse left him alone while they organised a bed.


How long then passed? The fireflies danced. The parrots took to shooting around the room with their wings tucked in. Charley Pickastick walked around doubled over, squelching water between his toes, dripping it from under his armpits. Later, much later, MacCracken confessed to himself that he had sipped more tincture of opium than was wise. His dreams were over-vivid. Life declared itself to possess an unbearable contradiction. He remembered butterflies with wings of patterned lace, each with a thorax of blue flame and a singing voice of beautiful simplicity. More like a hum. In Spanish, too.

He also remembered this: his jaw being taken in a capable hand and twisted slightly, so that it clicked with a sour piercing pain. That was surely no Nurse Parkington. His head had seemed cradled on a firm swelling breast, not a voluminous one. He had wanted to lie there until eternity dimmed. There was a perfume of orange blossom in the air. There were eyes, and they were dusky green as midsummer night's eve.

When next MacCracken woke he found his jaw bandaged and his ankle firmly taped. He darted his glance around. Who had it been?

'Theodora was here,' said Mrs Covington, giving his hand a squeeze. 'She will be back later to tend you again, unless Nurse Parkington comes in the meantime—but I assure you, sir, Dorry is capable.' Mrs Covington leaned close. 'Only don't call her Dorry in front of Mr Covington. He don't like it. It's not the *Spanish* way.' She acknowledged MacCracken's condition with a laugh. 'Not that you'll be saying a word to anyone.'


'Did you see her?' the Afghans asked when they called to say their final farewells that afternoon, bearing a tray of yellow-eyed mullet for MacCracken's supper, that he would be able to eat only after it was mashed into gruel and fed through his lips with a narrow spoon.

*See who?* He raised his eyebrows.

'Theodora,' they said. 'The woman in patterned skirts and a flowery shawl who made a bee-line for the Covingtons' cottage. Her skin was white,' they dropped their voices, 'her eyes emerald, her reddish hair a miracle of careless calculation. It was tied in a single braid and tossed forward over her shoulder. Her shapely feet were encased in

black slippers with silver stars, and is she an actress? We
don't know. Apparently, yes. Couldn't tell. We are rather
struck. Goodbye!'

'By the way,' said Mrs Covington when they had gone,
'your friend the bookseller, Mr Evans, left a package. He
said I was to give it to Mr Covington, and so I will leave it
here on the shelf, where he keeps his library. And Miss
Ferris came with a note.'

*Pass me the note*, MacCracken groaned, whereas he
more than wanted the book, he craved it.

He opened the envelope, peered at its contents, and
abruptly folded it over again. A stab of pain more acute
than the physical came from the words. MacCracken had
read such accusations of aggrieved expectation from a
woman before. He disliked the feelings they stirred and had
never expected to feel them again. But that was the pattern
of his life, it seemed—to go round and back on himself. The
last time was in Boston when an attack of difficult breathing occurred every time he considered certain promises he
had made, which he had breached for his own sanity,
leaving a broken heart and an insistent, hostile attorney.

But this from his quiet Georgina? The easiest, most malleable trader of animal caresses it had ever been his fortune
to enjoy? He pressed the letter in his fist until it crackled,
and then he dropped it to the floor. When had he ever said
he would marry her? Hers was a hard blow following
Covington's punch; just as unexpected; but not as promising at all. She appeared to believe that 'Dorry' was his
secret lover. One or other of the light-witted Afghans had
put the thought in her head, no doubt—and why?—because
of a streak of humorous malice poisoning its way through
MacCracken's friendships, he guessed. Lightness of character was a subtle poison. Sometimes MacCracken's friends
didn't like him the way he didn't like himself, and the
punishment exacted might almost have been chosen by
himself.

Was he a mere seducer, as she said? He bridled at the lady's choice of words. She was no stranger to delight. And believe him, there was something useful about someone—MacCracken himself—who ignored your opinion of yourself as a not so young woman any more, and devised a picture of you in their mind, and then led you towards that picture until it became you, and you liked it, and before long there was a craving in you for its consolations, because without them you found you were nothing at all, nobody. Was *that* a seducer? It was how Miss Georgina Ferris first became MacCracken's night visitor, slipping through a gate at the side of her father's house and coming through sheltering trees and arriving at his window. She was a woman of thirty-two years and of good family—daughter to an ailing sea captain, who had obliged her to make him her first priority. She had few obligations after dark, and so MacCracken sequestered her under his roof. She had spoken a thousand times of her gratitude to him, a younger man. But now this. 'Viper. Double dealer. Weed,' when all they had wanted was an easy time on a chaise longue with a glass of champagne and a tickling whisper in their ears. When all *he* had wanted at this end of the earth was his pleasure taken to the limit in his own snug cottage, without having the bother of getting to Sydney and seeking it there among the would-be marriageables with their narrow hopes. And why not? He believed it was his bargain with posterity to seek love, and have it to hand if he could. Marriage itself wanted no less, but MacCracken had a sworn dislike of the marriage vows, as such, and Miss Georgina likewise, he believed—until this note.

**M**acCracken became aware of Covington standing at the door in the half light. 'The folk over the way are grieving,' he said. 'Nought is of comfort to them. They wail and gnash their teeth and throw themselves on the ground. *Look* at me, MacCracken.' The doctor-become-patient turned his eyes sideways at the old sea-dog. 'You never look at me and see me true.' Then, to MacCracken's amazement, Covington began to undress himself. His clothes rustled with sand. Shedding it from his fingers he stood in his underdrawers brushing his grizzly chest-hairs. 'The soil is precious insubstantial in this part of the country, being nothin' but grit. Mrs Covington has great success in growing thistles for her milker, that's the best I can say for it.' With a puzzled, irritated manner he slapped at himself more—muttering 'thistles'—and then peeled off his last item of clothing, his damp, clinging long johns smelling distinctly wet-horsy: until he was revealed stark naked.

The reminder of horse fitted him any way MacCracken chose. His lips articulated themselves around words disdainfully, as if avoiding prickles. His yellowed toenails curled slightly upwards demanding a trimming, and it seemed as he rocked on the balls of his feet they were broad hooves. He was a bruised, scarred, ruptured dombey, this strong-boned, coarse-sinewed man, smoothly round in the

belly with a maze of subcutaneous veins standing out. He
brushed his flattened pubic hair and peered downwards,
lifting aside a squat, mottled penis. 'Get out from under
there you shallock.'

There was no falseness about him. Nothing about him
detracting from his dignity. And yet, it seemed to
MacCracken, it was on this point that Covington was in
constant battle with himself. Thus he had lain on
MacCracken's impromptu dissecting table at the very
start—a promising old cadaver with skin the colour of
plums—and emanated, even in unconsciousness, a disgrun-
tled questioning purpose. Today his colour was silvery, and
he was touched with a faint, rosy finish from his exertions—strawberry roan you might say. 'Tip him over and
straighten him out, wrestle him down till he ceases to kick,'
thought MacCracken, mentally lugging him to his dissecting table for another try. Cleave him from head to tail and
what would you find? Not the thing that Covington was
craving, surely, not the longed-for fullness of soul, not
there—only blood and breath of man and beast. For that
other part a different sort of examination was called for,
with no sharp instruments, lancets, fleam-toothed saws,
bandages or lint; but an art of understanding akin to revelation.

What would MacCracken have to be for that?

The question defeated him. Better. That was all. Better
and striving.

Mrs Covington entered, giving a quick, exasperated *tsk*,
and asked if Covington was ready for his bath.

'Does it look as if I am?' he preened.

'I am sure the doctor has seen grander sights.'

Covington stood slapping his chest as his wife busied
herself bringing his buckets of water. He made no move to
help as she strained and grunted, hauling pails from the
kitchen cauldron. Her service was a point he made about
their life together. In a while the tub steamed and was

near-full. When the temperature was to his liking Covington dismissed her and lowered himself into the water. MacCracken heard him sigh.

Was MacCracken to lie on the chart table until kingdom come, with a hard pillow under his neck and a light cotton wrap thrown over him? It seemed he was—to stare at plaster rosettes in the ceiling, cultivate his aching jaw and think his own thoughts as Covington played with the soap and cleaned his ears and nasal passages. Water slapped the sides of the tub, and before MacCracken registered that Covington spoke, he was aware that the man had been speaking for a time.

'... which is how it seems to me. Are you afraid of me, MacCracken?'

*No.*

'Respect me?'

*Yes.*

'Like me?'

*Not sure.*

Like? That was a flimsy emotion to have around such a monument.

'I don't give a damn if you like me, whatever your answer may be. But you have cause to be grateful to me. I fancy to think it is why you accepted my hospitality, for once without cavil,' Covington chuckled, 'even if I struck you.'

The question Covington didn't ask was whether MacCracken loved him—as he might love a father, say.

*Love you? How could I not?*

MacCracken had a sensation of warmth and affection riding above every other contradictory feeling and taking him over. Yes, he loved the man plain and simple—mere *liking* was below that peak like a flea on a plain. His own father, an insurance broker, was a sarcastic, disdainful man who undercut his son's achievements lest they highlight the disappointments in his own. Nothing was ever good enough for that crabby Bostonian, Sandy MacCracken, and on the day of MacCracken's graduation as a physician his Pa spent a good part of the celebration supper winkling from the assembled gowned professors the information that his son had scraped through his examinations on a wing and a prayer—and they had scant hope for him except as a provincial sawbones. So the hard-won scroll and mortar board that MacCracken had scored in the first place as much to please his old man as for any other reason, became just a fresh target of derision. Nothing was ever good enough for the man and so MacCracken gave up the bother.

Covington's bathwater slapped against the sides of the tub and MacCracken drowsed. The opiate suffused him with a kind of wisdom. Down the walls he watched the afternoon light turn pale as clouds raced over. Wind played around the cottage. He made Covington into the one who believed in him and bothered him for his own good, made him the one who passed back understanding of the world like a flaming torch, and placed it into his trembling hands.

Lying on the chart table possessed by hallucination, MacCracken felt himself swallowed and heaved out again by a whale, and it was not an unpleasant experience. A Covington kind of a whale he believed it should be called, all crusted with molluscs and trailing weed, and splashing the sides of its bath. He heard an insistent tapping sound and believed it was the whale's jaws clicking appreciatively and reminding him how lucky he was. There came that humming voice again, and low vowels of Spanish. MacCracken seemed to be swimming for an eternity in deep blue water made of billowing muslin curtains. Then he blinked and saw it was a set of trim fingernails making the tapping sound, and looking up into a pair of sleepy eyes, at lips that were half smiling, half downturned in seriousness, he hoped she was real. Those eyes at the edge of creation. That light drifting across them, a net of stars. Was she real? *Theodora?* MacCracken's character declared itself and broke from its flimsy framework. He would be strong—just by sighting an aquiline nose, pale freckled cheeks and a floating, encouraging smile. Theodora. He fell in love with her then and forever, and in a surge of resolution decided she would be his wife.

Covington spoke to her in Spanish, and then she was gone.

When next MacCracken woke, a wild wind was blowing. It was the southerly buster that rarely touched them, but now moaned around chimney pots and rattled a lonely pane. It slipped under the door and rustled the sand that Covington had spilled on the bare boards. Mrs Covington came in and chased the sand with a broom. 'What a mournful day we are having,' she said, and backed out of the room having a little pile follow her.

'What is it to be liked?' Covington stirred himself in his bathwater.

*I am liked. It has its moments*, attempted MacCracken.

'It is a great thing to be liked, next down from love. Or up, depending on the outcome.'

Was Covington asking about himself? About Theodora? If MacCracken fancied her one way or the other? So soon? He rubbed his sore eyes and pressed the tender upper part of his cheeks, where his sinuses were aching. He got the hint that matchmaking was in the air, yet thought it might be his own construction on the matter, or else it was of such a clumsy sort he could scarcely credit the process.

'That boy was liked, but he drowned. He drowned because he loved our liking.'

At the thought of Charley Pickastick, MacCracken felt his chest thicken. What an easygoing keen-eyed humorous young chipper he was. Involuntarily a sob made its way towards MacCracken's throat and he choked it back, fearing the pain to his jaw and regretting throwing his skinny watcher a silver coin the first time he ever did, thus starting the urchin's passion for diving from the wharf.

'If you're thinking he drowned happy you're a fool. It was a life ended, all its days stolen.'

'I'm a fool,' thought MacCracken sourly. 'Will I never get it right, even when my thoughts are being read, old bully-boy?'

'What is it like to have love taken from you? Given so freely, and then taken?' Covington raised himself from the tub, cascading water, and reached for a cloth to dry himself. All his sinews were taut and he twisted himself around as if ready for another dive into the deeps. 'Do you know that, MacCracken? Can you tell me?' Covington advanced on the temporary bed with the cloth around him like a Roman toga. 'These questions gnaw. Why do we begin asking them, and when we do, why is it God hides his face?'

*Don't stand so close to me dripping grey water, your breath smelling of green waterweed belches, old timer.*

'We are better off as dogs. They live on liking, loyalty, and a few odd bones. They have no knowledge of death, but only their eyes grow sleepy, and they dream.'

Covington eyed the crumpled letter that had fallen to the floor.

'What is this?'

*My dignity, if you please,* objected MacCracken, his voice box making only a muffled squeak.

'How does it start?' grunted Covington, echoing himself and smoothing out the crumpled pages and eyeing the words greedily.

*Don't know. And if you are alluding to Miss Georgina Ferris and her disappointments of me I would ask you to mind your manners, sir.*

'This letter ... '

*Yesss?*

'It is most satisfactory ... '

*You think so?*

'Except it shows you in a flimsy light.'

*Indeed.*

'You shall improve on yourself, MacCracken.'

*I had the same thought.*

'Otherwise there's no hope for us.'

*Us?*

Darwin's new book, that was of such importance to the man, and by connection of almost unbearable curiosity to MacCracken, sat on the shelf untouched, wrapped in Evans's waxy brown paper. Covington knew it was there and MacCracken knew that its presence ruled him. Shedding his toga, Covington's eyes swivelled into every corner of the room except the bookshelf. He pulled on trousers, shirt and smoking jacket. 'Where is my damned cravat?' he bellowed. Mrs Covington scuttled in with a hank of spotted silk steaming from the smoothing iron. Covington allowed her to fuss, to stand on her toes and work at his collar with a stickpin the size of a skewer.

'There was never such a strong handsome chump as this one,' she opined over her shoulder, making sure her husband couldn't read her lips, 'so cheerful and bright when I first knew him. Courageous, with the world at his feet. He needs you, doctor, don't let him down.'

Cheerful and bright were words MacCracken had never considered for Covington. They served the way lush and verdant did for a desert.

When the good wife was gone Covington stood rather pompously dandified with his back to the package while he filled his pipe and set up a blaze of tobacco smoke. His eyes looked puffy and tired. The exertions of the morning had done him no good, this man who had gone without sleep more nights than one.

'There is something I cannot do alone,' Covington declared.

*I already had that hint.*

'It concerns the matter of non-creation.'

MacCracken frowned to convey ignorance.

'*If* it is a fact,' Covington continued.

MacCracken still didn't understand. A tighter frown gathered between his eyes, which he shot at Covington like a bullet. *What in the name of blazes' brightness are you talking about?*

'Non-creation,' said Covington, 'is the idea that the Garden of Eden and the Flood of Noah are merely stories men tell each other. That when God said, "Let there be light," there was no great flash. There never was. Non-creation is the notion that there's been more time on earth than any mortal is able to calculate, and that those who love the Bible are fools. One such a fool stands before you, MacCracken.'

With this, Covington gave MacCracken such a hang-dog pathetic look that MacCracken made a gesture of offering his hand, which Covington, to his astonishment, came and took and squeezed.

*You a fool?* MacCracken answered somehow. *But you move me, my friend, almost to tears. No fool could ever do that.*

'There is something you don't know about me,' said Covington querulously. 'I have been shy to show it to you, lest I seem a relic like a mammoth or the sabre-toothed tiger. I am a natural religious man. I have my *Pilgrim's Progress* by rote, my Genesis, my Psalms, and my Gospels all in order.'

Covington mopped his eyes and blew his nose. He found a chair and took up a position at the edge of the table, where his eyes met MacCracken's at an equal level.

'You see, I once had a great text that comforted me, and I still have the same text, but it brings me only fear.'

*It goes?*

' "The things that are seen are temporal, but the things that are not seen are eternal." '

Covington let the text hang in the air, then added: 'I was put to work digging out the unseen, I played my part, and believe me, MacCracken, I was a great worker in the enterprise. Most willing, you might say.'

With that Covington made an inner decision, jumped from his chair and crossed to the shelf and grabbed the parcel in both hands. Without delay he tore the string, ripped the paper off, allowing it to clatter to the floor, and stood staring at a brown leather cover with a gold-embossed spine and various commonplace features: pages; edges; binding—a book was a book. But he did not open it and, as if it burned his fingers, took a long stride over to the chart table and planted the freshly-minted volume on MacCracken's chest.


MacCracken leafed through the pages. *On the Origin of Species by means of Natural Selection, or the Preservation of Favoured Races in the Struggle for Life.* It took only moments to see what was promised—a veritable Genesis of life. Despite blurred vision, an aching head and confused emotions, MacCracken wanted nothing more than to settle down for a marathon of study.

*A knife, a knife*, he mimed through his lips.

Covington cleared his throat.

'A name may be written in there. Doubtless in small print but writ there all the same, an acknowledgement of my existence.'

MacCracken made a sympathetic noise. *Would that be so terrible?*

'See, in my pride I crave it. I am moulded by the devil.'

Covington fetched a knife for the uncut folios, and after an interval of slicing and page-turning MacCracken reached a Register of Writers Referred to in the Text, a

Glossary of Scientific Terms, and an Index without any COV nestling above COWSLIP, where it would naturally belong.

*Your name is not there.* He gave his friend a beseeching, apologetic look.

'Then damn me for an imbecile,' exploded Covington, and strode from the room. But before getting very far he spun about and returned, looking everywhere at once, then reluctantly met MacCracken's eye.

'Look under birds.'

*Birds?*

'Finches.'

MacCracken brought a fingertip to the point of his tongue, leafed through pages, and looked under FI. He saw FEAR, FEET, FERTILITY, FIR-TREES and FISH, but nothing else.

*No finches.*

'No *Geospiza magnirostris?*'

*I'm impressed*, mimed MacCracken. *But no.*

'Are you certain, man?'

MacCracken made a grunt, *Look for yourself*, holding out the book as best he could under the circumstances.

Covington lurched towards the chart table seeming about to take up the offer. The most desired object in his eyes it seemed was the book, and also the most feared. He turned and left the room. The front door of the cottage banged fearsomely loud as he made his exit.

Silence. And then just the sound of the dying wind in the chimney space, and the voices of the women fussing at the other end of the house: 'What's he aggravatin' about, for pity? Perish the man and his tempers.'

MacCracken hefted the book to his chest and continued reading:

When on board HMS Beagle, *as a naturalist, I was much struck with certain facts in the distribution of the inhabitants of South America, and in the geological relations of the present to the past inhabitants of that continent. The facts seemed to me to throw some light on the origin of species—that mystery of mysteries, as it has been called by one of our greatest philosophers. On my return home, it occurred to me, in 1837, that something might perhaps be made out on this question by patiently accumulating and reflecting on all sorts of facts which could possibly have any bearing on it ...*

MacCracken read on. His concentration was absolute. His eyes swam under the spell of discovery and excited imagination. The letters on the page seemed edged by rainbow lines. And here was a word of thanks: *I much regret that want of space prevents my having the satisfaction of acknowledging the generous assistance which I have received ...* So there it was. Didn't such thanks, naming no names, embrace Covington chiselling away on his rocks at the edge of the ocean, collecting trinkets and putting them in tin-lined boxes for his benefactor? But no, on further reading, it was naturalists Darwin meant—a rarer breed than carthorses. Still, MacCracken did not have to read far to confirm that what he held in his hands was a distillation into sensible theory of the entire majesty of the world. Even reduced to theory the majesty remained inherent, the mystery still swirled. It appealed to MacCracken's temperament and excited him greatly.

His finger went to the tip of his tongue. Wetted, it returned to the corner of a page. Gripped the paper. Raised the page to a small curve. Rustle of the page turning, scraping over. A feeling of sinking through time as the next raft of words rose up. He felt part of time with a deep contentment and understanding—*all* times past and future as well as this very particular moment of his lying in this room, topped and tailed, waited upon, pampered, drugged and given the poetry of existence in such a form. It roused his admiration as an essayist to stand before a pile of such rich gleanings, and see them prodigiously exploited.

*We see beautiful co-adaptations everywhere and in every part of the organic world ... between the wood-pecker and mistletoe ... in the humblest parasite which clings to the hairs of a quadruped or feathers of a bird ... in the structure of the beetle which dives through the water ... in the plumed seed which is wafted by the gentlest breeze ...*

'And between the stumbling old carthorse and a light-headed fool as they circle each other closer,' thought MacCracken—beginning a pattern of replies and counter-explications that were to be his companions for a good time now, as he made Darwin's book into his own. *Artificial* selection; *natural* selection; the two phrases rocked back and forth in a symmetry of understanding.

The artificiality and contrivance of MacCracken's circumstances were not lost on him. Calculation and counter-calculation were involved in the whole long day and ran back through the two years of his friendship with Covington. They always had a plan for each other and reached towards it in their dealings. That was the artificial side. The natural element was in the instinct that kept them glued. Which led where?

Further and more rapid dimming of light ... Sweet odour of oil ... Lamplighting ... Fuss ... MacCracken was a willing partner in it whatever it was ... The clatter of Mrs Covington and Nurse Parkington getting a bed ready in the next room ... Conflicting moods cutting across: the thrill of the unknown in Theodora; the sour reality in his use of misunderstood Georgina ... The dimpled pudding with raisins for eyes that was Mrs Covington:

'Up you get. Lean on my arm ... '

Nurse Parkington on the other side, leaking liniment

through her pores as if she drank the stuff ... The limping
of MacCracken through into the small room matching his
own side room in 'Villa Rosa' ... The settling of him into
his new bed, a made-up couch ... His profuse thanks as he
was fitted as comfortable as two stout, strong and wonder-
fully good housekeeping women could devise ... His sheets
a froth of Egyptian cotton—easy as a cloud to lie on ... His
twisted ankle kept free of the pressure of bedclothes by a
wicker basket ... A lamp moved closer to him ... Objections
made:

'Do be sensible. You must not read now, doctor, you
must rest.'

*Just for a while.*

'He'll want to know,' nodded Mrs Covington.

'Know what?' shafted the Nurse.

'A distressing nonsense,' said Mrs Covington, twisting
her hands in each other. 'He relies on the doctor, see ... '

'Oh. Until *he* comes back, then,' relented Nurse, adjusting the lamp.

'If he comes back at all, Nurse Parkington,' said Mrs
Covington, going to the window and peering out into the
dusk.

'You mustn't fret.'

'How can I not?'

More mutterings made about the thoughtlessness, the
outrageousness of Covington. More defence of him by his
long-suffering wife ... Laudanum drops administered ... A
bedpan proffered. Then MacCracken left alone again and
the book taken up again. *The mutual relation of beings ...
The wide range of some beings, the narrow range of others
... Domestication. Artificial selection ... Variation under
domestication ...*

MacCracken began injecting Covington's voice into some
of the propositions before him:

*Crossing in the wild gives only limited variety.*

That spoke of humility. The fierce little tribe of Covingtons clinging to steep ridges and guarding their cattle among the dingo dogs.

*Huge variety possible under domestication.*

That spoke of ambition—and a very present one too for a man of property and clumsy pretensions. Why else did the Covingtons of this world take their wild-bred daughters to Government House balls, and try getting them joined with officers who were, back through the Home Counties and spa towns of England, in turn bred for such conjunctions, being adapted to getting rich colonial wives?

There was a trace of perfume in the room where MacCracken lay. Memory of a smile hovered at the edge of his understanding. Had the fleeting Theodora rested there, changed her clothes, sprawled on the imperial couch fanning her pale forehead while MacCracken hallucinated? By any reasonable standard of feeling he should be free of desire in his sickbed. But selection came in three forms, he read by now: artificial, natural, and then there was sexual, and the last-named was stronger than death.

MacCracken skimmed. Leafed back. Darwin's story was set far below man in the order of things, in the vegetable garden, the worm patch, the bird's nest, the frog pond. South American landscape and buried, ancient bones played an important part. Yet although there were no men, as such, treated in these pages, hints of them were everywhere—in every potent old sire of various broods, in males' dedication of life to the continuation of the line—in the draught stallion, the bull elephant, the oxen and the rhinoceros with its lowered horns and armour-plating. MacCracken lifted the sheets, peered down his trunk, and studied himself in the raw. He'd never quite thought of the penis ennobled as such, but there it was, ever-hopeful, the flagpole of life's claim on itself.

So what *about* MacCracken and his common urges

under the sheets? When it came to the attributes of sexual selection, which was he—the peacock with the showiest feathers, the gamecock with the sharpest spur, the lion with the boldest mane? None of it fitted. He had only his impertinence, his rather lazy-faced intelligent features, to recommend him. That and a fullness of desire making him ardent to his sweethearts, the several women whose hearts he'd crossed in life.

MacCracken understood what was being said against the comforting and the familiar in Darwin's pages—about the seething profligacy of creation and its object to breed at any cost. Because only custom created nature in the sense of what was proper. Otherwise wildness was all.

MacCracken allowed himself a whim of rapaciousness. Chasing Theodora through a grassy shade he caught her and had her on a muddy bank. The illusion was decently succeeded by a picture of them on the chair-deck of a steamer on the Italian lakes, smoking cigarillos and planning the architecture of a nursery wing.

He dreamed a while longer, then put his aching head into the book again. It was a great challenge to understanding.

Nature, wrote the hero of the *Beagle Journal*—now raised to greater heights—must have its universal laws and they were surely as obvious in the wild as they were in the stars. Because if in one place, why not in another? Laws were accepted in physics and astronomy, otherwise the arrangement of the heavens would collapse. An apple would never fall into your hands from a tree, nor a wave advance along a shore. The work of the wave and the seeding of the apple were well known to common sense. Lineal descent, likewise, was accepted in the farmyard and understood by the simplest of observers. Darwin wrote that he consulted farmers and breeders and they gave their thoughts. So domestication in Darwin's thinking was a demonstration of longer-term processes in the wild. Selective breeders summoned into life whatever forms and moulds they pleased. They imagined a spotted dog or a heavy-haunched draughthorse and within a few animal generations brought them into being. When Darwin looked farther afield—into the shipload of specimens the *Beagle* brought back—he saw that naturalists needed to learn the same lesson. He ventured confidently to look back thousands upon *thousands* of generations. Therefore all naturalists should understand how species in a state of nature were descendants of other species—though few did, it seemed, and stayed rooted to the idea Darwin shared before he began his study, which was that God made each being fit for its station in the world, and allowed it to vary only in small ways, as the diet and climate of its station dictated. How magnificent though was Darwin's idea of the wild, thought MacCracken, where forms were chalked out upon a wall of stupendous extent, not by men, but as it were by a greater force, most powerful, and utterly invisible except in effect.

In such a way MacCracken sketched out his thoughts of Theodora.

Now MacCracken was weary and let the volume slide from his fingers. His mind went to those green eyes again. Rested in them. Dreamed of attachment. How many shutter-exposures had his eidetic memory made as he lay stirring hazily, and Theodora took him in charge, leaning over him, expertly twitching him in the region of the zygomatic fossa, and so putting his jaw in place? A picture he retained was of her face turned half-away, chin slightly lifted, fiery hair freed and cascading, eyebrows memorably arched and somewhat thick, as if they underlined a style of thought. A sadness in her downturned mouth caused his heart to yield its independence absolutely. He believed it was for the simple reason that she was absent in that look, and he would never be able to reach her where she was, that his longing was found measureless in the very moment it began. Then he remembered. She had been carrying a lamp, holding it high to check the room on leaving. There she was, all of her—a small woman, well proportioned, intense in her fine-boned perfection, and youthful in looks although perhaps edging towards thirty years of age. Whatever, his heart went still.

The optical impression revived with hallucinatory clarity, MacCracken took it forward in time, playing with it in his imagination. Nothing became impossible, then, and all was found. Theodora spread her patterned skirts. MacCracken placed his head in her lap. She stroked his hair, twisting a lock in her fingers and giving it a tug. She laughed, although he had no memory of her laugh at all. They smiled into each other's eyes as they reclined on grass beside a fast-flowing brook. The stream was under Mount Monadnock, in New Hampshire, where he'd gone fishing as a boy. The glassiness, the twisting currents and ceaseless motion of the stream had always held him. His father flicked his rod upstream and MacCracken declined to join him, rather sulked, stressing an independence of being and a withholdingness in his nature.

A rubble of shells and small stones lay under the flow. Nothing much seemed to happen as the torrent passed over, but then one of the pebbles or shells leapt, jerked, and changed position. Whole numbers of them gathered and piled, broke away, ran in pairs, bunches, and continued alone, or else were all gone and disappeared in a cloud of debris and mud as part of the stream-bottom went. Finally the water was clear again, and it seemed as if nothing had ever been any different. The sun shone, the dew glistened, and life was ready for whatever happened next.

# BOOK

# On an Ark of Creation

# 1833

It was Mr Beskey in Rio, where Covington had first gone to find advantage in working up his ambitions, who sparked him to thinking of natural history as lucrative. Beskey spoke of names in Holland, Germany and France before selling him his insect collection in a broken box to use as a guide. But blow Covington down when he came to learn the rules from his Derbyshire gent—because half Darwin's procedures broke such rules when they started together, and letters followed them from England complaining they hadn't done it right; beetles mouldered in pill boxes; spirits leaked from jars; birds got mingled because the young master was confused between a lark, a pigeon, and a snipe—bewildered by which *circle of creation* or *inosculation* they matched. Covington was to overtake his employer in respect of a certain sense of order, and was often praised for it, too. Except his precision in the matter came to be the burden he carried through the world.

Beskey had inspired Covington to go all out, to tackle the luxury trade, the amateur cabinet, as well as to stroke the purse of serious scientific endeavour. Covington always remembered what was needed. Leadbeater's taxidermy agency in London was one place to send; the Maison Verreaux in Paris another. The demand was for knowledgeable collecting, trained individuals, good shots and exacting shots. Covington had never fired a gun as yet, had only stared at one and carried another, and knew a toucan from a sparrow, a Brazilian lizard from a serpent, and that was about all. He was merely the essence of ambition, that day when Darwin took him on, pestering the gent on their bark and insinuating himself with him as hard as he could. Meantime he absorbed his tricks from Beskey:

'Write down on card where you are, the time, the date, the place, and be full about it because it gives *cachet* to your find—and your client, be he never there himself, wishes to feel he were. And what if it is a new discovery? *You* may not know, but *he* shall be wild to prove it.'

So the next step to the state Covington was in had been starting his own bird collection. In South America aged seventeen years he devised a way of shipping specimens out under his master's nose. It was done through the services of a merchant's house in Buenos Ayres, owned by a certain Mr Lumb. Covington knew how such businesses worked: if you wanted something smartly done, and little talked about, you went to the chief of clerks and made their friendship. Understand he was loyal, always, and would not consider undercutting what he was employed to do, and for fair wages besides—but sought private booty when there was room for it in a bird-basket, and gave his spirit every chance.

Over the year they sailed south from Rio and La Plata to Tierra del Fuego, and then came back again making a zigzag survey of the coasts, their *Beagle* going like a diving duck at the wind and storm. It was the worst weather in the world when they made those forays, nosing away below the mirrored cove where Covington had travelled as a boy, in past Wollaston Island and Nassau Bay. Seas black, wind raw, timbers wet, faces pink, lips chapped, sails ripping, stays torn asunder. After sighting Cape Horn it took them three weeks to sail thirty miles. All were afraid of the vessel broaching-to, but their Capt was a full Christian and kept his faith.

So at last they came into sheltered waters, where even if wind raged one hundred miles an hour in the outer roads the bay was never disturbed, trees hung over the water, and reflections were true. There they put Revd Matthews ashore with all his furniture and crockery and his civilised Fuegians. Covington stared at the one he had grabbed that time off Brazil and wondered what had struck him, because now that he was deep in employment his impulses were not so humorous any more, and Miss Basket was already getting her finery torn, her face poked with curiosity by her savage cousins, who hardly knew her, for she had forgotten most of their tongue. She was on her downward slide and Covington was on his up.

After a short cruise they returned and took Revd
Matthews off again lest his eyebrows and furze bush were
all plucked out by reverted savages wielding mussel shells
as snippers. The mission to the natives that had filled
Capt's thoughts for a very long time was a wreck. The
*Beagle* sailed back to Patagonia and Monte Video like
the Grand Old Duke of York a-wheeling and about-
turning. Revd Matthews was not such a smart aleck around
the deck any more and mostly stood in one position,
clutching the lower ratlines, letting the spray flick on his
face and his shoes fill with water.


During a foray on land Covington was taught how to
shoot. They took a walk in sandhills, looking for something
to pot. Covington made a welter of loading the weapons,
putting too much powder in one, selecting wrong-sized ball
for another, and being fortunate when he fired that Mr
Darwin wasn't standing in front of him, because he let-off
before he was ready, and scorched a hole in the air. Luckily
the game was plentiful and not over-shy, and they walked
on a bit over the low, featureless country, crested a rise and
saw a flock of small deer, and so pulled down under the
crest and readied themselves.

'What did you say to get this work?' Darwin quizzed.

'That I was willing.'

'Aye, that you were willing. You rubbed hard enough
with that. But I mean with the guns, Covington, what did
you boast to me, that you were a fine shot?'

'I never told a lie.'

'Then come out with it true.'

'I said I had a way with guns.'

'Well, was *that* a lie?'

'Not if you consider the truth of it now.'

'I am lost. What is the truth of it now?'

'You are with me and correcting my mistakes,' Covington sulked.

'So *I* was to be your opening and your way with guns?'

'I think you are,' Covington nodded, risking his new employment with a flash of brazen confidence. He gave his powder horn a tip. Powder ran down the side of the barrel and Darwin berated him. It was the wrong sort. A finer powder was better with a ball the size they needed, as its faster action threw the ball farther. Every deliberate clumsiness Covington made, Darwin came back at him with a correction and an instruction. The rags Covington used as wads, to seal over the top of the shot, Darwin said were contemptible. They showed inferior musketry and many on the ship could have put him right, but he was too proud to ask. (In fact he had asked John Phipps and had not got very far, because given his choice Phipps never used guns but took to nets and snares, lime twigs and bells, and amused the ship's company by carrying a songbird on his shoulder.) Darwin would never use rags, cotton or tow as others did, because they were tinder-like and might leave a spark in the barrel. Crumbled leaves or grass would answer at a pinch. Darwin himself had occasionally, in a desperate hurry, loaded and *killed* without any wadding.

Thus prepared they sneaked to the top of the rise again, and the deer were even closer than before, complacently ripping at the coarse herbage. But in a moment they took fright, and Darwin swung his rifle with them and brought the leading buck down to its knees. Covington ran out and cut its throat. Darwin smiled to see him topping the rise like an ancient Briton returning with the kill, noble in forest and noble in glen, as he said, the buck slung around his neck—a vision of undaunted perseverance and tamed savagery in the service of civilisation. Back at the beach Covington confidently skinned the deer, butchery being a skill he had in plenty, chopped it into portions, and it was cooked on a fire of grass and weeds, dry stalks, sea-wrack and a few planks washed from the dim ocean.

Before they returned to the *Beagle* Covington took the guns out again, and potted a hare weighing twenty pounds. Taking details for the Game Book, Mr Wickham was snide with him, affecting to believe a sailor paid over to a gent was no loss to a ship's company at all—even though a feast was made of the meat and the gun room declared it the best tasting ever. Capt had himself a new clerk to make his weather entries, a Mr Hellyer, and boasted he was the neatest hand the bark had known and a determined amateur of natural history as well, a bird man par excellence. This Covington knew for a hurtful stab. Mr Hellyer looked as if he hardly knew what a bird man was but would make himself one if that is what it took to win a captain's heart. He could not even swim.

From then on the *Beagle* was no longer Covington's sworn bark by Capt's own decree. Covington was without duties aboard her save as they related to supplying his gent with his needs. When they were on the water, riding the waves—shifting to their next anchorage and survey place—it was all examining, pickling, packing, labelling and stowing. When that was done it was sewing buttons on shirts, drying clothes and damp papers on the coals, filling lamps with oil, cleaning, tidying, killing cockroaches and lice, rearranging books and looking into them when the chance was given, and getting a mite of intelligence sparked. Covington's fine copperplate was hungry for words. No matter how tired he was at the end of the day he turned to a task that was to continue, then, into the unseeable future—the copying-out of Darwin's notebooks, journals and important letters.

Riding the pampas with this man he called his Don C.D. he was not so bound to set duty. Always letting his horse go a little ahead. Always wanting to do just a little better than the one who taught him most of what he knew. Fashioning a noose from the stem of an ostrich feather and riding after a partridge. Galloping around in tight circles gaucho-style until the partridge looked at him stupefied and allowed itself to be caught. Such displays of spirits leaving the other one cold—the young naturalist looking overlarge on his pony, dropping his chin to his vest, lifting his eyes and studying the horizon more intently and his nag's ears all at once.

*Have you quite finished? Let us press on.*

Around the camp fire Darwin was often wordless in the face of Covington's cheeriness. He was cursed by an inability to unbend—to make idle observations when needed, as the riffraff did, oiling their existence with gabble. Nothing would come out, and so a willing boy made the running with ease—'Would they have fine weather tomorrow?' 'Would the ship return on time?' 'Are they guanacos looking at us from the dark?'—a patter of nervous hooves—'Aye, guanacos—they ain't Indians at all.'

Darwin's lack of effort dinned in his ears. 'I daresay we should.' 'I suppose it must.' 'There are no Indians nearby, I have it on authority of the posta commander.'

Why the gross prickliness with Covington a Darwin would be hard-put to say. *It was akin to bullying and must be a reflex of some kind because it defied the will and excited peripheral nerves. Was murder at the same root of instinct? Undoubtedly.* He would later use such words. But he would have to be honest some time, spitting out what it was. Servant or no, he didn't quite like him. It was enough to make the angels weep.

'Red sky in the morning, shepherd's warning,' said Covington as a matter of habit. 'Wind before seven, dry by eleven.'

'It is *always* dry hereabouts,' said Darwin, scooping a handful of dust. Covington saw he was making a resolve, pursing his pinkish, wind-chapped lips; rubbing a sandy, sun-burnt eyelid—he was going to have to be more patient with his manservant if this was to be possible at all. He knew that a sailor lived by the weather, and Covington was a sailor almost from childhood, was he not, having weather in his bones? No wonder he tested it constantly, boringly, tediously. Darwin understood that. Besides Covington came from a survey ship. Their very duty led them into places that all other ships avoided and their safety depended on being prepared for the worst. Darwin could be fair-minded, no mistake. So that Covington carrying a certain over-readiness from the bark to the campo was to be expected. And yet his damnable chatter! And the sore irritation! And the relentless presence of the lad! Must he hoist his crotch, spread his legs way wide, and jig a little from the knees every time he readied himself for bed?

'Every night we reef our topsails,' Covington chortled when drawing his blanket up to his chin and giving it a shake, 'so as to lose no time if a breeze forces us to move.'

'I think the worst we shall have tonight is a dew,' invariably observed Darwin—at which Covington laughed immoderately, and agreed they well might.

*

Covington came back from shooting:

'Covington!'

'Yoi?'

'You've got me far too many skins,' said Darwin. 'More than I asked—'

'Look, I have sorted them.'

'You have indeed. But Covington—'

'Yea? What have I saved you?'

'It must be a week, I confess, at the least.'

'There you are, then.'

'But you have too many of the same kind. What is this?'

'It came skimmin' along the water.'

'But it is half shot away.'

'A pinch is better than a peck, ain't that right?'

'Not in the case of birds, I am afraid. It is a lot less than so. This might be a scissor-bird. I cannot tell. Its beak is shattered.'

'I used a light shot,' said Covington defensively.

'If you could get me a scissor-bird—'

'What then?'

'Well,' said Darwin with a certain false heartiness, 'I might write to the king for a commendation. It is a great rarity.'

'Then a scissor-bird it will be,' Covington swore, but thought, 'I am not slow, except to condemn; I am not witless, except when I do not know, and cannot keep up with Cambridge men; I am not a child, who will be satisfied by excitement. So what does he mean, when he condescends? That I should go away, hide my face, hobble my tongue, and only come back when there is *hefting* to be done?'

Neither was Darwin comfortable with Covington's fiddling a few scrawled tunes in the pampas night, making a monstrous interruption to his thoughts, which were deep and slow. 'Enough of that caterwauling there,' he would declaim, pretending to be in the most hectic good humour about the thing. Then he'd turn his back and pull his blanket around his ears, cherishing his sleep on dry land in contrast to that at sea, where there were always footsteps on deck, thumps, clangs, bells, voices against one's ear in the most private of cubbies, and always the queasy motion of the waves.

Once or twice of this was all it took, and Covington crept away from the fire where he sat propped against his saddle-bed looking up at the stars, plunking the strings of his Polly Pochette with the flat of his hand, the battered instrument hard against his ear, making a repetitive medley collected along the way, a music touching his mood of obedience and making it rankle a little—telling of dark eyes, red roses, flashing steel and spilt blood. '*Lua na testy munha*,' he sang, remembering a rough night of his life. *Moon, my only friend.*

It was a marriage of convenience they had, and Darwin was like the fiancée who gives her consent to the match for reasons of suitability but through lack of love rues the intimacy—yet all the time lauding the practicality. Darwin's beau ideal of a collector when they began was not Covington but the coxswain and captain of the maintop—Covington's dear friend, the walking cyclopaedia of God's mysteries who could set bird traps as prettily as steer the cutter, rig a sail or produce a Biblical reference in his best Bunyanesque style. Darwin owed a debt to Phipps for releasing Covington from a closeness of personal friendship without hostile jealousy—a dissenting, somewhat conspiratorial and devotional closeness, it would seem from the way people talked about it. From Phipps's and Covington's point of view it was a comfortableness so close as to seem like carrying home with them wherever they went. This attachment was famous in the ship for having a whiff of inseparability about it—there had been grumblings that the two were as tight as wax, that there would be trouble from Phipps if Covington was removed from his sphere, and so on. But no such trouble came. Phipps had a vanity he could draw the gent to their circle of prayer, which wasn't so fanciful, when you considered the number of times Darwin 'amened' through Sunday service. It was Phipps had the good effect on people, they were soothed by something in his heart—immovable-seeming, yet lightsome in the practice of his faith, which he sniffed as a rare breeze, and went nimbly pursuing to catch with his texts all arrayed like sails.

Covington was not such an inspirer. Or say he inspired something else.

**B**lood was a matter Darwin came back to roundabout more times than he knew—by the light of their cookfire under the stars, and when they were stretched easy on the ground and sucking their pipes and looking for something to say—blood and back to blood, the blood of base men, as he put it somewhat forcibly. Blood by which we mean lusty thoughts.

'What—of rogues, do you mean?' asked Covington, wishing to help his gent out. Darwin was such a great cloud of confusion on divers occasions.

'All classes, I suppose. The usual run of men.'

He doubtless meant Covington himself, reference to a whoremongering reputation based on the crew's japes over his encounter with Leza the Uglificate in Bahia, and with his Hickory of St Helena—even going back to the time he leapt in the water and grappled the Patogonian maid with her small *pechos bezoomos* and lost eyes, her *ojos amoratodos*. Nothing was secret in a ship, all was known in relation to others. But the stories had become bent with time—his leaping was upon that maid, it was said, whereas it had been around, and they had touched in surprise, not by foul design.

The gent was no obvious gossip, and never so bold as to ask outright. But his need was clear enough on what he wanted to know, and Covington interpreted him thus: Who were the lustiest Beagles aboard, and how did they argue a woman to lie with them, by what signals did they devise the conjunction, and did the men believe there was a way of knowing a maid was willing, as distinct from other maids, say, as they were seen making their promenades in Buenos Ayres, all those pretty *senoritas*?

'As easy as knowing if a squirrel is in the humour, you mean?'

'Customs, routines, mechanisms, devices. The whole gamut if you will.'

'Oh, he would ease up to her and very soon know,' said Covington.

'By what means, however?'

'It is much the same as in the fowlyard—by means of her response to bravado and show. Your Rio Plato miss,' Covington airily declared, 'will blush like a rose-bush and feign indifference, but it is good for a man to be hot-blooded, as they say, *de malas pulgas*, fire strikes fire. They think our Englishness is very strange and call us *budin blancos*.'

'White puddings?'

'Something of the sort.' In fact he had listened to soldiers at a military post using the word and seen them point their thumbs at the gent himself, a touch unfairly. 'About what you are saying ... '

'Pray continue.'

'There is not much to add. In other cases, among the common sort, why, it is the same the world over for a sailor, he speaks his intent and shows his coin.'

'There is no great mystery there,' said the gent.

'Or shows his coin first,' continued Covington, 'by way of flirtation.'

'Really?' asked the gent, carefully flat.

The subject was over. Darwin could never have a conversation with Covington except when he wanted to know something. Now he knew. It was never easy.

It bothered Darwin that Covington came after him with perfect judgement as to when to shatter his thinking— Covington knew that, but couldn't help himself—calling out, pestering to be told something he would certainly never remember and was irrelevant to the task at hand. Then the Don's thoughts would take wing and flee irrecoverably away. *Volcanic plains: beds of coal: lakes of nitre and the Lord only knows what more.* What was that about? There were times early on when they might have struck each other.

'You have explained the hemoffrodiate,' Covington boomed one day, 'but never got round to saying if it was the same with people as hemoffrodiates—might they cross with themselves one time one way, another way next time?'

'Give birth, you mean?'

'Aye, to themselves,' guffawed Covington.

'We are not a *hermaphrodite* race of people.'

'Not on our bark, we ain't, for it's a hanging crime, for sure.'

It was the closest the two of them came to ever wondering about Phipps in that way. If his leading of boys was mixed with a longing to spear their bodies along with their souls, then it was kept well hidden, even from Covington.


In his science, around Darwin, Covington was too much like Swift's footman in a smart district of town who tried to learn all the new-fashioned words and oaths, and songs and scraps of plays that his memory could hold. Covington did it from the lexicon of Cuvier they carried aboard. He remembered best when inspired or passionate. If tired, bored, out of humour, then it was futile to expect remembrance of what a cognisant naturalist or a well-versed country man would know without question. At other times, in their beginning together, he thought it advantageous to *think* like his master, or how he thought his master thought, making connections and not just reporting facts but whatever facts might seem to be important. How full of himself he was, all spunk and spittle.

'What was it you saw flying?'

'It was very like an ostrich.'

'How far off the ground did it go?'

'It come over my head.'

'Then why do you say ostrich? You know they don't fly.'

'In the way it threw its wings and ran along the ground.'

'Pray tell me what you saw.'

'It was a carrion vulture.'

Field ornithology led the way to systematic and descriptive ornithology in the normal course of events, but there was no such expectation around Covington. The getting-and-gutting school was the only one he needed to join, but still he struggled with his terms, hating his limits, going against them. Covington was best with birds, as time went on, and that was where Darwin himself was weakest. So they complemented each other somewhat. One reached for feathers while the other hammered rocks. It was a matter of getting into Covington's head that limits were not just inevitable—no two persons on earth having equal capacities—but indeed desirable and useful, if properly worked. It was a matter of beating Covington into the ground.

Invited to sweep the horizons, Covington returned festooned with plenty, and would persist unto death from exhaustion, if need be, preparing skins and packing specimens. Thus despite their personal irritations Darwin was enabled to increase the sphere of his observation as if he had three or even four assistants and not merely one.

Covington knew his worth but still felt the emptiness between them. It was his gathering small hurt. Darwin complained there were problems with being a naturalist in the field, separated by thousands of miles from fellow naturalists. He always sadly wanted somebody to talk to about his finds but knew in his heart there were few who cared one pin about them. Seaman Covington? Wouldn't he do? Hardly at all. Not a serious idea. He was the ballocky parish bull, Cobby the young 'un, the gross leaper as he was characterised on the bark—a specimen of vigour and open-faced potency breathing in a man's face overmuch. So much indeed that Darwin refined his manner of keeping him at arm's length without spoiling his use. None of this was lost on Covington, who might have no science but could read the emotions the way naturalists read their Carolus Linnaeus. Darwin's technique with his underling was more suited to making a plan or sketch in botany or geology than in human dealings. Regrettable, but there you were. Covington was to be valued as a phenomenon, but it couldn't make him liked, and the matter of his popularity among his shipmates only emphasised the distance between them.

It was a great mark against Darwin that Covington understood all this and more, and that Darwin never saw it, or if he did, considered it of no importance, absolutely not.

And if only breathiness, sly grins, closeness, over-enthusiasm and an air of cunning were all. But the damnable stirring of the blood that went with it gave Darwin hot flushes—he *blushed*, spouting crimson like a Roman candle when he least wanted to, and so the most reprehensible of all situations came to the boil, and he found he was bonded to a servant who looked at him without moving an eye-muscle, yet seemed, ingratiatingly, to wink. It was too uncomfortable altogether.

In this Covington had his own back. He vented his blood. The outcome most strangely was a child. Theodora.

All the while FitzRoy's chart of South America was added with detail. It grew, multiplied and defined itself to the eye incrementally—headland by headland, bay by bay, reef by hachure-marked reef and island spiderwebbed and caterpillared with ink lines, depth-marked with numerals carefully calculated, cross-checked from soundings, and notated with warnings of rip-tides and places where safe anchorage might be found for the world's fleets.

Likewise Covington's boyhood had shaded into its new life. Away from Darwin and his manner of always making him feel unfinished and rough, he was a wandering taxidermist in the best of styles. His kit was three pairs of scissors, a whetstone, a whalebone needle, and spring forceps made of brass. He looked at his former shipmates and saw how few of them were contented at all. Like him they were caught young and broken-in before they reached their years of discretion. That was how the navy held them. But unlike him they weren't old John Phipps's boys any of them. They didn't have good faith. They were rarely happy so far from home, and when it came to their devotions they only spouted them in peril at sea, swearing they would live blameless lives if saved from drowning.

They made a short cruise to the Falkland Islands and Covington took a long walk with the guns. Darwin asked him to keep an eye open for reptiles and limestone and report back if he found any, but all he saw was a low undulating place with stony peaks and a brown, wiry grass growing on peat. There were no trees, therefore few birds, only snipes and waterfowl. The best he got was a brace of rabbits. Phipps won the day capturing a live fox and it was taken aboard.

A man came running with news that he had found some clothes and a gun on the coast. Thick strands of kelp swayed among rocks, rising and falling in the tide with melancholy precision. In a short time there was a darker lump visible in the kelp, and when boat-hooks were found and the mass hauled closer they saw it was pale Mr Hellyer tangled in it. With some difficulty he was disengaged. He had shot his bird, and while wading to take it his legs had entangled and he was drowned. A pestilence was in the lads' moods as the *Beagle* sailed north to Patagonia waters and made its way south again, then turned and repeated the punishment. It occurred to Covington that it was strong business he was about with his Don. Hellyer in contest for birds was gone. Wee Volunteer Musters wielding the gent's gun was gone (having died in Brazil). Joey Middleton, so ardent, so fine, was rigged in his hammock, given his balls of shot for weight and dropped to the ocean floor, and whose service had he sworn himself to betimes? Covington's fingers were bleached and dimpled like his master's from spirits of wine as he pickled mice and shrews, spiders, moths, and made sure there were eyeballs staring out from inside every glass. Did they want him for their companion?

There were times when Covington looked around, saw their severe Capt, and imagined himself on the deck with old Noah looking for the rainbow marking the end of the deluge, and being first to sight the dove with the olive branch, announcing the finding of dry land. 'The Scilly Light ahoy!' Nothing could touch his optimism then.

The Great Flood of the Bible was spouted on their travels as a favourite story. The idea of a ship with animals aboard was a picture of cosy creation. The Ark in everyone's mind wasn't much bigger than their *Beagle*. It had the same goats and chickens in stalls, the same tar, mouldy wet boards, damp wool and stiff tarpaulins, frozen ropes and water washing the decks to the knees. It had the same kind of Capt, precise in his rule and close to God. The midshipmen and officers all had nature-enthusiasm, and their own insect collections, but nothing like the ones Covington went around making under the guidance of his gent. Nor anything so particular as Covington kept for himself, either. The gentry were all the sons of Noah in their way, keeping a little back for what came after. As for the crew—who had been life and breath to the boy until now—they understood the parable if they bothered with it at all, but saw the whole enterprise of collecting specimens as gentry-business mostly. They went on feeling excitement as they found it, and when they did not—creatures of the moment—only grumbled. They had a regular gripe over their Cobby getting above himself and being bound to the Derbyshire gent. It had been heard before about him in different guises. They wanted a cut in rum whenever he returned from shore, meaning a share of his rations, and called him a land-swab, which he greatly resented. Old John Phipps showed restraint on the matter. He was a friend of Darwin, now—did fowling for him betimes, and coddled and kept alive the ship's fox. Phipps had the parable to work himself into as well. After Noah came to Ararat and emptied the Ark, the width of the world lay open to him. The animals went forth and multiplied. The beached Ark was dismantled and used to build houses. So the old tub that had once been everything shrank to other uses in God's imagination, and the whole wide world became his vessel.

The captain made Fuller his servant after the death of Hellyer. He was a friend of Harper the sailmaker, and formed an addition to the band of devotees—now making four—who were led through their catechisms by coxswain J. Phipps. Not that Covington was back on board so very often, but when he came—roaring with boasts and with ostrich-shit in his hair—there was no slackening in his devotions. He slapped a good handshake with Phipps and they thanked God they were both in the land of the living. Then they went to their duties and met when they could, and had words both quaint and pertinent. Covington was over struggling with faith and had it inside himself secure as a well-knotted reefer. As for Phipps, he would never change. He was a rock.

It was a good enough kick Covington was on but it was like him to want more. After pursuing Darwin's favour so hard at the first, and beginning to know him well, the question occurred to him whether he loved his gent or might ever love him at all in the way he needed, with a full flowing heart. He liked his master's good-humoured energy when they were busy together, when Darwin forgot his affability with inferiors—but that was only his surface, and underneath rumbled something large, and it was like indifference to men. He would have what he wanted and stay comfortable with the getting of it, and would often forget to say thanks. Lucky for him, then, that Covington was at one with his faith, and every time he saw a cross—a twig, a mast, a crucifix—he said his prayers and his burdens fell off him, as Bunyan's book said, into the mouth of the sepulchre.

*But by the blood of Christ*, he resolved all the same— *take a master who cannot love and you get a servant unable to love either*, pudet pigetque mihi, *and all of a gent's Latin he cares to spout*, de verus, mirabile, *and so on.* Fine words were as dust on the ground, as Job said.

But what matter—it was good work that he had—and a humble bee in a cow turd thinks himself a king.

It was October and spring. His work was killing small birds. He made his way through grasslands to the south and west of Buenos Ayres travelling like a pilgrim with everything on a packhorse and equipped with his gent's best gun and bird-baskets. For each green paroquet and misto finch he took—every lapwing, cow-bird, tyrant-bird, humming-bird and swallow—he took another: making two collections, the larger for his gent, the smaller for himself. Using mustard-shot and dust-shot that barely parted feathers and peppered skins, creating little damage except for a heart-stopping dunt, his long-tapered bird gun resembled a blowpipe with its elegant stock. When fired it cast a smoky breath in destruction—and more protracted than any breath came the report in Covington's ears, a ringing echo lasting long afterwards.

His Spanish came out like a barking dog's, and although he dressed as a horseman in the fashion of the country, in poncho and loose pantaloons, he climbed from his mount and walked more often than he rode. His packhorse followed carrying gear and extra guns, shot-pouches, powder-flasks, caps, flints, wadding and balls. His destination was an Estancia Thompson where there was rumoured English hospitality, a family of boys, and a governess by the name of Mrs Bonnie FitzGerald. A family of boys was as good as a wide net spread over a countryside—what they did not bring in, blinking in their handkerchiefs or peering from their shirtpockets, was not there to be found. Covington carried a sketch-map in his pocket and letters of introduction. Darwin rode a separate route hundreds of miles distant. They were well away from each other after being too much together. They were to meet in Buenos Ayres in a month's time, when the *Beagle* would come for them. So Covington was free to collect what he chose under broad instructions.

He was happy as never before. Falls of rain hung in curtains across the horizon. Sheens of water lay in plates and narrows, and putting his horses to the jog he splashed through them. It was a countryside of rolling green, but threadbare, like a cloth left out in the sun too long. Everything was budding and fresh but there was a roughness, a patchiness under the green compared with spring in England. Mr Thompson's name was given by the Buenos Ayres merchant Lumb in a letter of introduction: '*His estate is the longest settled, and I believe handsome and large.*' 'And I believe you have never been there, Merchant Lumb,' thought Covington with a grin. There was *large* in that country but *handsome* was choice. He would see what it was when he found it and know its style when he reached there. There were too many hovels and too many betrayals in South America for anyone to be impressed in advance. Lumb had failed to supply Darwin his shot and powder as ordered because of a prohibition under law. It was due to the war against Indians that was running. They were being killed, each man, woman and child, by General Rosas and his murdering henchmen—chased from their shelters and cut down to save ammunition and left to die in the open. In that district the war was already over and bones were under every tree. There was a saying that when the Indians came the vultures would have a feast. In one place a corpse hung with its skin dried. It had bird-scarers, pendants and mirrors hanging off it and making a sharp unpleasant sound. It was agreed by landholders that the extermination was needed. Proof of it was there were no arguments over cattle any more, they had the free range, and Rosas was the greatest cattle man of all, owning four hundred thousand.

Covington was not so shy of soldiers' rules as Merchant Lumb and got his powder by bribery and persuasive charm. So maybe it was that a poor savage lived another day through the loss of some favoured bird, *avis rarissima*, as the gentry said. It was what Covington told himself as he went along in his work, carrying mail in his letter-pouch including a missive from a Rosas man, one Colonel de las Carreras, to Mrs Bonnie FitzGerald.

Covington had bettered Lumb on a small matter by obtaining a box of shipping tags ideal for labelling birds, and not paying a single *reale* for them. It was done by a wink to the chief of clerks who minded the storeroom, and who promised to forward his birds to Leadbeater's. With a string on the left-end the tags could be tied to swing clear of a dead bird's legs, but not be loose enough to tangle with others when the birds were laid away. The tag did well for anything, from a hawk to a humming-bird. In no way were the birds he prepared for Leadbeater's given preference over Darwin's; except the satisfaction Covington felt was greater. It paid to attend to such things as tags and so win praise, one small perfection on the part of a servant counting for many in the eyes of a master. In this the young Darwin was no different from all the masters of England— a good part blind, another part hopeful, and a last part condescending. Also somewhat deceived.

It was Covington's caution, in company with his gent, to stow away his best accoutrements—embroidered shirts, showy boots, silver-buckled belts and daggers—and not play the exquisite too much. In this he followed the mood and manner of his master, whose vanity resided within— except for a bashfulness over showing his profile, owing to a nose like a lump of pear-wood nailed to a door. For his Don was a plain-faced man with little interest in wearables, save they were sturdy and cheap, such as the black duck trowsers he bought Covington on his account, and the nankeen cotton shirts they wore at sea, that itched and rubbed with dust on the trail. Once they took a bag of sweepings from a granary and Covington crawled around a splintery wooden floor with a brush and pan. They sent seeds home to England to be sprouted, to discover if they were changed since they first came from Europe. Darwin said: '*We* shall be a Botanical Problem when we reach England, Covington, it will be curious to find if we are changed to be like the men here, if so, I will answer for it we shan't be much improved.'

It was good being trusted to go shooting on his own. Covington's faith was strong in the day. Birds sprang up before him as he went along. They flickered along the trackside, mazed through trees, and ran along twigs to get a better view of him. It took fine skill to drop a warbler skipping about a tree-top. To stop a pit-pit rising from the grass was even better. Most of his targets held still when taken, though. They were God's innocents and not gun-shy at all. With his muzzle-loaders always ready there was no hesitation in him. After his early greed he aimed to one side and risked losing the bird rather than blast it to perdition in a flurry of feathers. After downing a specimen he basketed it away, reloaded, and went on. How could he say what it was, that he thought, or rather had no need to think, in the spirit of how he was? A perfect gratitude warmed him as good as the sun. He marvelled at daily wonders.

Birds were scarcer on the plain, prolific in shelter, thick in places where men never ventured at all. He'd been at it long enough to know their ways. Birds lived in poisonous swamps, on icy peaks in the southern winds of the Horn, where sails were stripped but not feathers. They inhabited every station of existence, some of them scuttling through leaves like rats, others ringing the world on narrow wings and never alighting on earth. Birds outrode storms at sea that sent strong men to their last account.

When hawks flew in Covington's way—and other birds of prey—he learned it was best to wait for them to find him. Their shadows steadied, and he looked up, and that was their end.

Owls he scared from bushes in daytime, and at dusk potted by rubbing wet lucifer matches on the foresight to make a pinpoint glow, tracing them in their utter silence until the quiet was shattered. To gather owls' corpses in his fingers was to feel a spirit sift through them. There were no softer feathers known.

At his prayers he asked God's forgiveness for harming his creatures, and knew, as sure as if the answer came back to him in person—like it did to the old prophets—that God was pleased with his labours.

The birds he prepared were stuffed into sleek bullet-like shapes and narrowed towards the tail and feet. There was no need to pretty them up. They were not made as objects for a show case, but for a naturalist's examination cabinet. Yet was there anything more beautiful than a simple feather overlain with another, so detailed and tough as a rasp? Covington mused half-sleeping that birds' column-shapes were expressions of music—tubes for wind that he could hear through his beginning deafness—the sweet and the loud in rows with their heads tilted slightly over, their beaks aligned and their feet twiglike and tied with shipping tags. He stroked the dead birds feeling their unearthly lightness. He made a final settling of feathers as he put them down, using a palm-up action of the hand rather like cupping water. There was a text that ran: *Death is not welcome to nature, though by it we pass out of this world into glory*. It underscored him.

With every specimen he ever packed away in its bed of cotton he hung over it this moment of saying farewell; he dressed the feathers that were well dressed before; perfected every curve; finished caressingly and put his bird tenderly down, as he hoped to be shriven himself when his time came.

Rumour went ahead of him that he was English and strange, because he killed what others left alone, walked and led his horses more often than he rode. The country people, the gauchos whose paths he crossed—and who gave him their hospitality at night—expressed their pride in never walking at all. They used their feet only in hobbling to their saddle-beds, hopping to their fires of dry thistles for roasting meats and brewing their yerba mattee teas. They used legs for jumping from thistle-clumps, cutting throats of Indians or each other as it pleased them. Covington's safety was up to their whims. He sensed their dangerous humours but left such decisions to God, and laughed in their faces. They were like cripples or were bred to be centaurs, you might think, living on horseback in their fresh-cut cowhide boots crusted with blood, with silver spurs and spaces leaving the first two toes dirty and bare. Such toes were made for gripping in stirrups, not dirt.

The watery October season gave him ducks, waders and water-fowl aplenty. Their differences of paddling and coping with water were always of interest. The knack of shooting loons and grebes was to aim at the water in front of them, because they did not go under just where they floated; as Darwin had observed when teaching Covington to shoot, they kicked up behind like jumping jacks and plunged forward. It was a keen pleasure for Darwin to see cleverness in nature, and Covington learned the habit from him. The steamer goose used its wings to flapper along the water; the ostrich raised its wings like sails, and went up with its helm. Mostly it was the small and the overlooked that Darwin wanted collecting—except for bones, which might be huge, heavy and petrified.

So Covington with his boxes, spirit bottles, fly-nets and guns was after a year the match of his master in this regard—a meticulous plunderer.

C oming to a river, wide and overflowing its banks onto grass, Covington stripped and washed in the chilly current. Then, stepping naked from the water, he was surprised by boys throwing stones from a stand of willows. When he was cleaned and dressed in his dry clothes he sneaked round behind them through bulrushes and bailed them up with his fists. They were defiant and spoke English. He said that he was English too—at which they were all over him as if they had been awaiting a signal, looking into his bags and whistling in admiration over his knife and gun and his shrivelled talisman of puma tongue. They said they had heard of a traveller coming along that might be him, 'A man who knows everything,' and so they yabbered between themselves in Spanish, coming back at Covington calling him '*Il naturalista*', at which Covington smiled and shook his head: 'Nay, that is someone else. My master, you mean.' The boys chorused their offers: Did he want birds' eggs and frogs? Lizards? Skunks? Owls? *Anything* in that line? Covington said yea, he wanted an armadillo brought to him live so as to have the joy of sticking it through.

The boys ran ahead of him sending their spotted dog in wide circles scaring up game, and in quick motion disappeared over the horizon. Covington forded the stream—and three miles distant on the other side, across the green plain like an enchanted island, he gained sight of the Estancia Thompson. It stood at the end of a long muddy track in a field of poplars. A flash of window-glass gleamed as he came closer. The boys rose from a hollow and looked back to be sure Covington followed and understood he was to link himself to their play. They were like small birds darting and swooping and then diving between a gap in the trees and disappearing. Covington longed for company and followed at the jog. His saddlebags flapped, his packhorse scented clover and lolloped out to one side and ahead. A sack of tin dishes and cutlery made a racket, but his specimens were well secured and he gave his saddlehorse a jab with the spurs to get him cantering, too.

A short time later Covington pulled the horses up to the trees and stood on an embankment. A double row of poplars blocked his view farther in. Their balsamic odour came as clean as any sea smell. At the edge of his hearing came a sound of a surf running up a shingle: leaves shivered; and looking up at the clouds moving he had the sensation of standing on a deck. It piqued him with a memory of danger and a starless night when their bark rode without anchor and close to a Tierra del Fuego shore. There had been a great whispering of surf on shingle and a booming of swell on cliffs. They believed their Capt went close to mad that night with giving his orders, contradicting them every time the wind changed.

Then Covington scanned the foreground and noted that someone was moving through the trees towards him and trying to stay undetected while aiming for the best possible view back at him. He thought it was the boys again. But it was Mrs FitzGerald, as he would soon learn, on one of her wild walks. And he would soon find himself captivated by her and bound to her in a way that resembled his own taking of game, in that it was a kind of death she dealt him, and through her he went to another kind of life.

Then he saw, sauntering towards him, a brown rat. It

had a piece of offal between its teeth. Recall, thought Covington, his gent and his catechism: 'Look for variations among common types.' Here was one plain to see in the form of a common pelted rodent almost gingery in hue. It crept from the broad ditch running along the front of the poplars. Such ditches were dug to hold water for a dry season, and in defence against Indians. This one was a haven of burrows and a dump for bones, and a muddy obstacle for visitors to the estancia.

A sharp squeal and the rat found itself wriggling under Covington's boot. He bent down and twisted its neck with a ready jolt. Then he knelt and skinned the rodent with his narrowest boning knife and rolled the pelt around a stick. The action was leisurely-seeming but swift. A green eye, that he was acutely aware of, but scarcely knew watched, considered him the whole time with an iris narrowing and widening as clouds crossed the sun.

Coming back from the Estancia Thompson the boys were all forgotten and he rode with Mrs Bonnie FitzGerald at his side. It was a great trust to be a woman's lone guardian in a country where a traveller's entire safety depended on a companion. He would give his life for her and throw himself at her feet. That was decided in a moment, in her first glance his way. She wore flowing skirts and a straw hat with a black ribbon over the top, which was tied under her chin, and he was full of himself as her escort. Her hair was the colour of flaring flames. At night their hosts were the rough and ready who served meat, yerba mattee tea and nothing much more, and whose roofs of thistles and reeds barely kept out the rain. Each morning they were up at the same hour, in the faintest pearling of dawn. With the sun coming through they bid farewell to their hosts and were over the horizon at daylight and Covington was about his business. Throughout the day he worked under a wide sky following wings and submitting to instincts of nature. The land was another sort of sea. It was amazing how constantly on the move it was. Wagons, men, horses and cattle fractured the horizons, which in such flatlands were close. If there was a slight rise the world expanded in size hugely. A faint smudge in the morning resolved itself by noon into an ox-cart. Mrs FitzGerald was sociable and paid calls to passers-by, for she was known to several families. With some she might have gone ahead, but had attached herself to Covington heart and soul, as he was bearer of a letter of safe conduct from her husband-to-be, and she was glad to follow at his pace, and if she missed him in the morning she always found him again in the afternoon by stilling her horse and listening for the report of his gun. When eventually they met with Colonel de las Carreras he would be grateful to Covington, she said, and escort him into the city through enemy lines. She called him her boy. He didn't like the word, but the way she said it, with a warm breath in the end, and a look in his direction, made it all right.

Sheets of shallow water fringed by reeds and teeming with ducks lay upon the landscape. Here and there stood a flamingo, eyeing Covington with trusting stillness, and he never shot one, feeling himself ennobled by such stately presences, and besides, not wanting to carry anything so large. His trays measured thirty by eighteen inches, were four inches deep, and held three hundred and fifty birds six inches long. As a tray filled the drier birds put in first could be submitted to more pressure. A skin originally dried in good shape could later be pressed perfectly flat without damage. The only thing to avoid was contortion. It was like the work he did on the ship, he told Mrs FitzGerald— on the domestic side of his service, looking after his master in the fashion of a wife, the whole knack of packing birds corresponding to filling a sea-trunk solidly full of clothes, as could easily be done without damage even to an immaculate shirt-front.

The worst he did that day was shoot a heron, but it was from necessity, as he had resisted taking one before, and had an order for it. He was in love with the crazed crest of feathers above the creature's eyes that bristled when he took aim as if the bird by its stillness said, 'Dare.'

When Mrs FitzGerald wanted to know what all the taking was for, he explained about comparative anatomy and the men who made it their trade. They were Englishmen in London, in Cambridge, and in the country-side around where some were parsons with time for natural history. A few were Continentals, and the ship's library was heavy with books in French. But a man named Henslow was the chief of them at the Royal College of Surgeons in London. He was the leading bone man, and the target of all their digging. They had been prodigious diggers in the south before Covington was sent after birds. They had found the bones of a giant rhinoceros at Punta Alta, on the Patagonia coast. She did not believe him at all. Nay, he said, but it seemed that animals now living in Africa once roamed the pampas, and Mrs FitzGerald shivered when he said so, being delighted in her imagination for she had a fancy, one day, to ride pillion on an elephant under the tutelage of a maharaja.

At noon they stopped and spread lunch, and she had never, she said, 'had a finer, stronger, or handsomer young man at her disposal'. They ate salt beef, pickled onions, apples and pears, and tipped macaroons from a canister. He watched how she peeled her gloves from her fingers and put them aside. She saw him watching and smiled. She tossed sugar-almonds to him and he caught them in his teeth. She laughed, and every time she laughed she was prettier than before. A little plump, he thought, and too short by a half head, but desire mended deficiencies, and her emerald eyes hooked his hopes. Those eyes were hooded like a hawk's and she kept still watching him, till he asked what it was, and she said, with a pang, 'The ghosts of the past.' She had been widowed young, and was childless, and was tired of living in small rooms in other people's houses, always at the call of ignorant badly bred half-English children and finding herself regarded as little better than a kitchen slut. She had spanked a Thompson child and received her acquittal from service. They were not a civilised family. 'You saw the way they lived.' They boasted a glass window, a cushion, a fork and a spoon, and had a kitchen separate from the rest of the house instead of just an open fire in a smoky recess. That was their height of dash! She was half a mind determined to ask her Colonel to requisition their stock. That would show 'em. Mrs FitzGerald had not wanted to marry a countryman but Colonel de las Carreras wanted her very much, and would take her on her terms, and so she was going to make her life with him.

Mrs FitzGerald rested after lunch, lying on a blanket laid on a tarpaulin, a forearm over her eyes and her ankles crossed. Covington walked a short distance away, set traps, spread nets, and sat in the damp grass with a sack over his head watching plovers take worms. When the plovers came close enough for him to thread a finger into their feathers (or so it seemed, just by reaching out) he gave them a dusting of small shot and one of them fell down—only raising a wing when the breeze caught it—while the others scooped the pasture with their feet and took to the air.

He saw that Mrs FitzGerald was watching him, sitting up, her arms around her knees.

After another ten minutes he took the heron.

Then he was back to her and told her it was not a pretty sight to see gutted a bird of good size. She said she had seen worse and minded little. So he began his work on the heron, donning his apron, squatting on his heels, and taking the bird in his lap. Before she realised what the blade was doing an incision was made, and Covington was at his skinning. Worms and green slimy material came spilling from the innards, and lice from feathers crawled up his arms, disappearing into his hairs like an army on the run. He boasted he was able to make good small skins at the rate of fourteen an hour, a pace that impressed his Don, who had learned the art from a black man in Edinburgh. The quickest work Darwin ever did was eight an hour, or an average of seven and a half minutes apiece, and fairly good skins.

Mrs FitzGerald watched the skin coming off, the feathers tucking under and being hidden by gore. She was taken by Covington's boasting, and said with amusement in her voice: 'Do you improve on him in everything?'

'In practical matters if I can,' he replied, sniffing with a certain pride. 'Which is what he wants me for.'

He made a deft action with his hands, taking warm skin in his fingers and rolling it up.

'Oh, *that* is how you do it!'

He told her yes, it was the same way she ungloved her hands from the wrist, by turning the gloves inside-out to the fingertips.

She was pleased that he noticed what she did with her gloves. (They had pretty pearl buttons and were her pride.)

'Some people say, *pull* off the skin,' he said. 'I say never pull a bird's skin; *push* it off.'

'*Push* it off,' she repeated, and gave a low laugh.

Before she knew it the eyes were out, and she exclaimed that the eyeballs were much larger than they looked from the outside. She leaned on him to get a closer look.

'Are you faint?' he asked.

'No. I am interested.'

Next he cut out a squarish-shaped mass of bone and muscle. Pulling the neck back he brought the brains out, leaving the entire roof of the skull supported by a scaffolding of jaw-bone.

He had a mere pulpy mess in his hands. Or so it looked. The next motion was to bring the inverted skin back the right way.

They heard a distant whooping shout, and turned their heads.

Three cut-throats broke from trailing their cattle. They splashed over the plain with absurd curiosity and reared up, staring at Covington with his meagre, simple kit—

women's embroidery equipo, it must have seemed to
them—not knowing it was what the best taxidermists
worked with. They eyed Mrs FitzGerald, debating whether
her throat was white and succulent enough to slit and how
much bother would be involved if they had to get down
from their saddles to do it. Meantime Covington held the
mess of pulp in his hand, and they giggled and speculated
obscenities, stalking their horses around in a circle. Mrs
FitzGerald said nothing, barely acknowledging their cat-
calls, just showing haughtiness of eye, and then Covington
saw that she had a hammer pistol under her shawl.

Holding the long bill in his right hand, he made a cylin-
der of his left and coaxed the skin backward with a sort of
milking motion. It came easily enough until the final stage
of getting the head back into the skull-cap; this required a
bit of juggling; the watchers hissed; but he could not fail to
get the head in, because after all he'd got it out. When this
was fairly accomplished he had the pleasure of holding up
something that looked like a bird. It was slack and billowy
and yet still vivid. It was the white heron, or rather the
poor ghost of it.

The chief of their callers, a toothy gentleman, oozed
friendliness and courtesy, and issued an invitation for them
to stay at his estancia that night, it being well placed for
hospitality. His confreres grinned from their rotten gums
and stepped their horses back. Giving their crass farewells
they performed slashing motions like Turks, though they
had no swords and only their scarecrow fingers to indicate
what they would do to the gentleman and lady's enemies if
anyone harmed them before they took meat under their
roof.

It was hardly to be credited they were the same people who
gave out their hospitality that night, in courtly denial of
their murderousness. Yet the three welcomed them to their

hovel with waxed moustaches and low, craven bows, and later, after they had fed them as guests, made a grand production of parading their daughters. Then they looked serious and clicked their heels to the strains of a jig from Covington's Polly Pochette. In the dance the three came over and pinched his cheeks and giggled and kissed him. It was all the same blessed difference to them whether they loved him or not. Now they loved him. Now they loved him not. All was a whim, depending on absurd flashes of pride, supposed injury, false generosity, and woeful ignorance. They ushered their daughters forward but warned Covington if he impugned their honour by the merest flicker of interest it would mean the knife. So he praised the daughters' beauty and virtue but barely glanced at them as persons.

Around Mrs FitzGerald the men were fantastically fearful: they would not harm a stray hair in her exalted flaming head through knowledge of Colonel de las Carreras and his battalion, despite their adherence to a different faction of politics. This allegiance they greatly regretted, and might change sides if she spoke to him of their honour. They argued among themselves: 'No es justo juzgarla con tanto prejuicio.' (It would be wrong to judge her harshly.) She promised that she would do what she could with her Colonel. It was a country of cheap life, each taking his share according to his own estimation of what was due, knifing each other for little or nothing and killing Indians for the joy of the blade. They assured Covington of their protection of him. The women ushered Mrs FitzGerald through to bed. Now the men passed the bottle and called him compañero. With a spinning head he went to lie down.

At some time in the night, propped on his saddle for a pillow, Covington found himself facing an intruder. A hopeful face grinned and lisped, asking if it was true that he carried a lion's tongue in his birdskin purse. He admitted it was true. He showed what he had by candlelight.

It was like an enlarged chestnut, or the seed of a bean tree, with purple stippled roughness. The fellow, whose nose had its tip cut off, sniffed but dared not touch it, and crept away. Covington returned the object to a canvas sack he kept between his legs when he slept.

In the sack was his pistol and his letters from home, his coin bag, and a briarwood pipe bequeathed by his dead shipmate, Charlie Musters. The pipe had a silver band decorated with a pattern of oak leaves and acorns. When they sailed from Brazil they had left Musters behind on a rocky headland—dead of a fever caught snipe shooting. He had used the lightweight muzzle-loader of Darwin's that day and the next person to carry it out of all the ship's grieving company was Covington. 'I shall not spoil myself with mourning,' Covington had determined, 'as I did before.' In his death-fever Musters had raved himself into glorious battle, believing himself on the deck of a man-of-war splintering and burning. So he died in glory it might be said, and like a boy captain raised himself on a shaky thin elbow and appointed his belongings to various men. Covington did not know why the pipe came his way, except that he had once taken a puff from it, and judged it excellent, while Musters turned himself blue. Musters joined Joey Middleton among those never quite finished with life, never quite dead as they travelled the world with Covington, never quite alive either but cased in an ice-cold column of water that always needed warming. The pipe had a briarwood bowl like a roughened furnace where the tobacco smouldered and warmed his fingers. He smoked it on first waking and thought about Mrs FitzGerald in the next room. Then he tapped the bowl on his heel and gave a deep sigh, and thought of Mrs FitzGerald some more, and wondered if what he wanted with her would be the best thing in the world as he thought it might be, or if it would be even more, a nibble of heaven.

He had the cut-throats around him again in the dawn

light. They said they loved him as a brother, but really they wanted his boots. They were made by a cobbler in Monte Video to a pattern copied from a famous Spaniard's pair. He splayed his fingers inside them from pure pride, fitting them to his arm, giving them a buff of leather-soap, twirling them in the air and admiring the stitching, that was neat and regular, with a flourish of fleur-de-lis on the toecaps (that was supreme in the originals). Usually he kept the boots in a sack. But today he wore them, and his best shirt besides.

Covington and Mrs FitzGerald splashed over the edge of the green world. He had his killing to do, making deft work of stopping the hearts of small life. But he tried not to get his clothes too dirty. Mrs FitzGerald noticed the change in him. She teased him, and said he was *un grande galopeador* away from his ship. She didn't mind if he got himself dirty—wasn't that what men did? She went on in a vein of flirting and he boasted to her some more of what he did— grubbing up old bones and captivating new animals, smiling at the idea of himself so far removed from the lanes and fields of Bedfordshire. She made admissions of herself in return, being from a town in County Cork, Marlow by name, which she had left as a youngster riding pillion on a bay gelding with the son of an Irish lord, Mr Barry, who wished to marry her most devoutly but was forced home. So she was stranded, for a time, in Jamaica, until rescued, abandoned, then rescued again by Mr FitzGerald (who died), and found herself in Monte Video. When she spoke of her girlhood she removed her hat and loosened her hair, and in the heat of the day placed her jacket across the saddle and joined Covington in his shooting and preparation of specimens, taking an interest quite emphatic and useful.

'What is he like, your Mr Darwin?'

'He has more energy than me, and I have plenty for ten,' said Covington. 'If there is something he wants he will go to the ends of the earth to find it. He is clever, but you would not think so.'

'Then he is not like you,' she flattered.

'I am not clever, but I am better than him in ways.'

'I am sure of that. You keep telling me,' she smiled.

'You have heard me boasting. But see us together and you would have another impression. I seem quick beside him on some days. I have seen him stand while his shadow moves around him. He is all the time thinking, and he declares by a great authority that genius is patience.'

'Are you patient?'

'I cannot wait too long for what I want.'

'Does he rate himself so high?'

'No. You could say he tries to make himself invisible at times. But his effect is always strong.'

And then Covington let out an annoyed word, and said that he always ended feeling confused around Darwin when he expounded his science, as if his brain was stuffed like the insides of a specimen, with nothing but bark, leaves or cotton waste. Even FitzRoy their captain did not always keep up with him, either, though he tried, especially when the finds were special and exciting, and then he baited and mocked his ship's naturalist in the way these gentlemen used, that was foreign to the ordinary ship's company: 'I do not rejoice at your extraordinary and outrageous peregrinations, Darwin rarissimus, because I am envious—jealous—and extremely full of all uncharitableness. What will they think at home of Master Charles—"I do think he be gone mad",' etcetera, and words to that effect.

'One day I was jumping to understand what was behind what he did,' said Covington, 'because there is always that feeling, that he knows what it is and won't say. And I said to him, "Is it like those things that are seen are *temporal*, as the books say; but there is something even more *invisible*, and we might see that too, if we find *enough*," and Darwin was pleased with me, and said, "That is exactly right, Covington, full marks. The things that are not seen are eternal."'

'You make him sound like one of your English parsons with an answer to everything, and going about the countryside exposing the blessed mysteries.'

'He plans on getting a parish right enough, I've heard him say,' said Covington, 'you know, with a "good living" and "a thousand a year", it would suit him very well. And he longs to get his oats that way too.'

'His "oats"?' she smiled.

'Aye, cockeyed, ain't it?'

Mrs FitzGerald leaned across and placed her hand on Covington's wrist, and looked him in the eye. It was a way of reacting that made Covington feel important whatever he said, and he knew he could be boastful with her, or complaining, or saucy, and it was all the same. 'He wants a pretty wife on a couch in a vicarage, as he puts it, and I think that is where he beams his lusty thoughts,' he added.

'He has lusty thoughts?' she asked.

'Well, wouldn't he, being a man?'

'If he had been to the same school of advancement as you, Mr Sleepy-Eyes, I would say so.'

He thought it a pity she wasn't free, otherwise he would make his confession to her right off, that they might make a go of it, ride away, and be adventurers together. They went to their horses and started getting them ready.

'What is it like, going on pillion?' he asked.

'I was never so free,' she said, and without too much ceremony got up behind him, letting her saddle-horse follow, and he felt the grip of her arms around his waist and the pressure of her cheek against his back.

Then they were closer than speaking.


It was noontide and the next day. Covington was skilled at finding advantage in declivities and hollows, where he waited prone with his gun and lay back with clouds drifting over, watching larks climb almost from sight and then float down pouring out their song until they settled in grass near his head. Mrs FitzGerald lay beside him. He spread his nets and waited. The next lark that came he caught, and handed over, telling her not to put the least pressure on her fingers, or its heart would stop. She admired its gleaming eye and noted its palpitating breast, and then she set it free. Covington thought: If Colonel de las Carreras came upon them he would want to know what they were doing, so close and so smiling. Then he would take Covington as she took the lark, with ease, but whether he would set him free was a question. He would possibly shoot him or at the very best pluck his feathers. Contrariwise he might arrange for him to be feted a hero and be given a sash for his protection of a virtuous bride. You could never tell in this country what it would be—unless you were an Indian: If you were an Indian you would be chased, ridden down, betrayed, clubbed, knifed, slaughtered. If you were an Indian girl above fourteen years you might be kept and used, but otherwise Rosas's men went out in troops two hundred horses strong and crossed the plains to the snowcapped Andes. They went on chases five hundred miles long, night and day forsaking food and drink except for mares' blood. It was not wise to even begin thinking about it.

Mrs FitzGerald reached out a hand again, and tweaked Covington's ear. 'Did you hear what I said?' she asked. He hadn't, he was going deaf, and so she said louder, 'You have two cock birds in your bag and shouldn't you get you a hen?' Well, he told her, he tried to get one each of whatever he bagged, the cock and the hen, but was never sure what he scored. Sometimes he killed the same kinds over again without realising, skinning and preparing like mad, binding the trays to his packhorse frame and sending them off. Variety was what mattered—the differences showing in beak and tailfeathers, speckled backs, breast markings, often easily observed, sometimes harder to tell. Mrs FitzGerald seemed to follow what he said, but he thought she was not listening to the words as such.

They lay parallel, propped on their elbows. She moved from tweaking his ear to stroking it a little absent-mindedly. It was all about *phylogeny*, he said—what belonged with what, which bone fitted where, which gaps in knowledge an animal explained. If you had a fancy, continued Covington, you might think of birding, bugging, and bone-study as an argument running back and forth, in which the samples from the *Beagle*'s voyage played their part. When they got back to England there would be their great congregation on the matter, and Darwin would be in the driving seat.

'I am tired of your colleague,' she said.

'Colleague is flattering. He is a Cambridge University man and I am a boy from a boneyard. He is one of those who is always "joshing" all the time. But he never joshes me.'

'Is he rich?'

'Very.'

'Is he a looker, Syms Covington my lad?'

'Better than a pig and worse than a peacock.'

'Oh, he must be then,' she smiled. 'I never trust a man to tell me anything worth knowing about another man. It's in their blood.'

He liked it she called him a man after calling him a boy.

'*I* might tell you anything I wanted,' he said.

'What is stopping you? I heard you admired my gloves, but that is all. Do you have nothing left to say?'

'Nay, but you are betrothed.'

'Pledged but not bounden,' she said, a little tartly.

He took a breath and considered the danger of her. Eyes watched in these empty plains and made their reports. Who knew if any watched now? Troops were stationed not far ahead. There was a chance she would turn the tables and want him to flee with her if he made a grab for her bezooms, which he was close as his next breath to doing. Did he really want to run away with her? It was the whim he'd had. But not now. Not on his life.

Yet he drew close, and his blood was stirred.


Covington told it to Mrs FitzGerald as they rode along—that his master lacked warmth. Didn't he realise that if he wanted a dog, Covington would even learn to bark, and fetch a bone?

'You do all you can and more, and you have a great heart,' said Mrs FitzGerald, tilting her head to one side and looking meltingly at him. By unspoken agreement and mutual plotting they nudged their horses towards a grove of poplars that marked a slight rise on the plain. The ground would be dry in there and the place sheltered. A burnt-out chimney showed where a house had been. There were no other buildings around. As they came closer they could smell the sweetness of poplars.

The air was clover-fresh, pungent, tangy with new growth. Covington drank it as wine. Ribbons of smoke wafted through the stillness and hung in bronze bands across the lower sky. Covington believed he could sniff mischief-making gauchos a league distant and sense their savage guile. There were none too close, no smell of burning meats ribboning the flatlands. But if he was wrong he carried his pistol in his coat to make things equal. Likewise for ostriches, hawks, ducks, anything else that clawed the blue sky or pecked green grasses. He sensed where they were; went after them; and because he wanted them, he found them. A small dose of lead so as not to damage them, only kill them, understand, and then they were ripe for skinning.

In such a way Mrs FitzGerald was delivered to him in the grove of poplars in the warmth of the day. She stood down from the saddle before him peeling her gloves, and then impatient with the pearl buttons she tore them off.

For the rest it was soon done, a matter of a carnal greed and exchange of pleasure most satisfying and exhausting: an education in lasciviousness. It happened over two days yet seemed without limit and was rich with animal instinct. Their chiefest glades were swathes of old thistles, dark woody trunks from the previous season that only ever hid treacherous Indians (but they were gone) or wandering cattle (but they had no fear of them) and so they had only prickles to remind them of sanity, which state of mind was last in their thoughts. Their marks of civilisation were a coarse blanket, an embroidered pillow, and Covington's bird-hide with a length of cane holding it from the ground and making a shelter for two. With their game declared he was in a rage of wanting her any way he could. She held him off with a small hand, laughing and asking had he been with a woman before. His eyes hooked into her, 'Not a white 'un,' he breathed, and she cooled (he thought) considering his worth. *Not with a woman the way I want you, madam*, his mind rumbled, *either, with your legs in the air and your back in the dirty*. Would he ever get into her business-end this way? While she plucked at buttons and tugged pretty bows he tensed and waited. The bullock of Elstow had nothing on him, his blood so full that his eyes stretched tight in his skull. 'You are no help to a woman at all,' she taunted, but when he lunged, getting a hold of her bubs, she was exasperated and pushed him off. 'Not so fast,' and she asked him to kiss her but only to kiss her, nothing else. Was the horn in his trousers never to have its toss? She guided his touch, a quick hand for squeezing, a slow hand for stroking, and a chuckle around his getting to her under-silks—'*push* it back'—and a purr—'*roll* it back'. Her breath was sweeter than plums, all softness was in it, and when she fell back, crying, 'Take me, *buen mozo*, I love you,' in a most dramatic way, his need led him into her—quick, there, done.

And there she was with her skirts under her chin and her eyes like a cat's, and ready for scratching and hissing again. It was a good pattern they had.

Except that while they still had a day or two left of their ride Mrs FitzGerald began tidying herself too much, frowning with irritation at her jacket all stuck with burrs, and dismissive of Covington when with all his boy's humour he tickled her at the waist and grabbed her ankle to bring her down, and she turned on him hostile.

On their last night they were among friendly troops and she was queen. He was like a whipped spaniel over the quickness of her passion. Through the days before he'd downed everything on the wing in the neighbourhood, a figure of eighty birds and also twenty small quadrupeds, rats, mice, pests, night visitors. Uncountable more than that was the number shot and rejected as being spoiled. He was in bad spirits and totally gone on a rampage over his situation with her.

A ringing in his ears came on while he squatted on his haunches in a small mud hut, peeling birdskins by the light of an oil lamp, his forearms bloody and his fingers running with the warm, yellowish crop of seed-eaters. The business put him in mind of a useless meal. The sound came like a screech of wineglasses being hummed by a wet finger, and he looked through to the next dim room where Madam was dining with officers, to see if a game was being played, as he had seen the gentry doing at captain's dinners, but their cups were forged from clay as you might expect in that country, and they were eating meat from their knives. So he went to the outer door and stood watching the stars, wiping his hands on his apron. Maybe it was them—the heavens. They sang, did they not, as they rotated in their spheres? He was tired and his head was not right. Maybe it was time for three shining ones to come to him and salute

him with Peace Be To Thee, to set a mark on his forehead
and set a seal upon it, or something of the kind, because he
felt like death after enjoying her favours so mightily, and
death brought to mind sin and disappointment after such
ripe whirligigs. The sound came from within material
matter and from around it at the same time, which is what
he understood an emanation to be, and there was only an
eternal explanation for that. He was betrayed, confused,
stung, and reduced to a desperate nonentity and he wanted
to be gone from this life.

'Do you hear that noise, a small bell, a cricket?' he asked
her when she came through for air, sweeping somewhat
grandly in her new importance. He was conscious of over-
politeness to her, arch and offended.

'Dear oh me,' she hissed, 'whatever is wrong with you?
You are not in love with me, are you? Don't spoil it.'

Smiting his head with the flat of his palm he said, 'The
ringing, don't you hear?' Mrs FitzGerald looked alarmed
because he yelled. What sort of passionate fool had she
lumbered herself with, when she had only wanted his
desire? The soldiers that came through calling her back to
the table wondered at Covington crawling around on his
knees, bothering his head like a dog with a sore, looking
under casks and following into the corners of the room for
the source of the ringing. They blocked a hatch and closed
him in, and he yelled that he would have them shot, for
which they kept him longer, cackling as long as their
adored one allowed it, which was not for long, and then
spewing their apologies. A post-rider had been sent to fetch
Colonel de las Carreras and now shots were fired, shouts
were heard, horses whinnied being spurred most cruelly
and reined in at the same time to keep them frenzied. A
great jangling of silver was made and elaborate orders
shouted, responses given, and there went Mrs FitzGerald
open-armed to a portly, suspicious Carreras illumined by a
flaring torch of thistles carried by a peon.

The next day they rode into sight of Buenos Ayres and Covington had his last look at Mrs FitzGerald. A night's sleepless thinking had diminished her in his mind. She rode at the side of the Colonel resembling a pouter pigeon, and nothing would make you think there was anything in her policy except reflecting his importance. Carreras belonged to the Rosas party that was blockading the city and not letting food in or people out. He was a great Indian-killer no doubt, for Mrs FitzGerald said his cattle were all safe from them. The polished walnut stock of Covington's bird gun attracted the eye of Carreras and he wanted it, offering safe passage through his lines as his price. But Covington was astounded and refused, on the honour of his master, whose bird gun it was, and said his passage was already promised by Mrs FitzGerald, and that the Colonel was obliged on her honour to provide it. Carreras took this to be immense effrontery, which it was, being a correction of the morals of a man of pride, and also courageous in the extreme from an English weed, and so to be admired. He muttered to an assistant for a few dark moments, and then the assistant came over, bowed, and said he was Covington's guide.

They waited until nightfall. Mrs FitzGerald came and took his hand, and then kissed him, quite formally, on the cheeks. Carreras bade him good fortune. They went around the fires of one group of soldiers and skirted the fires of the next. Neither side was in a mood to fight. There was just a little skirmishing, a few killed by day, and by night a kind of peace in their war. Halfway to the town they went through a plain of mud and the horses went up to their bellies. It was where the gun slipped from Covington's hold. He turned around in the blackness and saw his escort leaning away from him in the floundering, as if adjusting something on the port side of his mount. They were skilled scoundrels all of them. Whether the guide relieved him of the gun or it fell in the mud he had his suspicions. When they reached the stones of the city he found himself alone. He learned that Darwin had waited at the Estancia of the Merchants and then given up on him, having caught the packet to Monte Video a few days previously.

So Covington was able to arrange for his private birds to be sent to Leadbeater's without other eyes around. He required payment to be made to Mrs Hewtson of Bedford on his behalf. It was a good feeling and went partway to restore his wrecked pride. At Merchant Lumb's everything was in chaos and there was constant fear of the town being ransacked. Covington waited on a balcony and saw the rebels enter, five thousand strong, and when he saw Carreras he ran down intending to face him over the gun: really there was a wish in his heart as hot and sharp as a dagger; but who should come along in a flurry but Mrs FitzGerald, almost stopping his breath, and she crimsoned and bustled him into an alcove and whispered, hot-worded, that he was the silliest, most callow fool, but she wanted him to remember her better, after all, and if he ever became rich from all his bird-plundering he was to write to her in care of Merchant Lumb's. Then she gave him a folded handkerchief, saying it was hers in Ireland as a girl, and he must treasure it close.

Her kiss was full passionate, dizzying warm, renewing every promise her touch had made, and she was gone.

Covington retreated inside himself somewhat as the voyage went on. From his close living together with Darwin he added patience and long-waiting to his virtues. He learned being one thing to one person, another to the next. It was a change enlarging his usefulness in service, though narrowing to his spirit. Or more truthful to say, packing his spirit down in the nether-hold until it cried to be taken out. He did not press so much for what he wanted any more. Silence bulged around him. Deafness boxed him in.

They sailed down the coast and he missed Mrs FitzGerald with a mighty sorrow that made him shed tears into her handkerchief. He did not know why, it was a painful souvenir, but he kept it till its scent faded, which was not long, then shoved it deep in his seabag, and took it out and stroked his thumb on it many times after. It was embroidered BF and had a four-leaf clover in green, and was a fine, soft piece of work. It was all his heart as he recollected his time with her.

The *Beagle* stood off-shore in a gentle swell. They were on the low coast of Patagonia, at Punta Alta near Bahia Blanca. Covington found himself standing on a beach where bits of worn skeleton rolled around in the waves.

They were old bones: all bigger than horses' and cows' bones by far, and strangely configured. To get them back to the ship's boat, at a riverbank landing, a packhorse was hired. Water and land intermingled dizzying a vision of one in the other, and the floating bark seemed to break from itself in globules of glass. Men broke in two walking only a short distance off, and the brain buzzed with excitement over what was found, the very brain itself breaking in half trying to understand what was in these problematical shapes that the living and the dead brought forward. The place was like a great slaughtering yard where hides only had been taken, and the rest left to rot, as was typical in that country (and how the Quentin House of Bedford got its profit). But who by, or by what agency had the bones been laid down? And what hides had they had, that must have been thick as turf on cottage roofs? All were agreed that a sweeping power of water must have brought the bones to the plains, and it was never cattle skeletons that came in such size, but debris of a race of animals gone from the earth before Noah.

Covington followed his gent's bootsteps scrambling a loose cliff and spent a day dislodging more bones still, making use of pinch bar and spade, thus proving his energy for ten masters and more. Darwin made notes and Covington asked him what he was about today. He answered, '*Quien Sabe?*' with a short laugh, but Covington insisted.

Gent rested his chin on the handle of his spade and said he was about getting good South American fossils to display in England, because there was only one, a fossilised sloth, in the whole Kingdom, which was held at the Royal College of Surgeons. It seemed what they'd stumbled on was unique, except in Madrid there was a *Megatherium* specimen described by Cuvier as having a curious osseous coat. Could these bones be it? For they put Gent in mind of living species of the armadillo, only bigger by far.

'I have trouble enough with your standard issue armadillo,' said Covington. 'The lead-bellied bastard even gets in my dreams, and I *still* can't pull him from the ground, owing to his strength.'

'See what you can do with this one,' said the gent.

'I shall, if it takes all day and night,' said Covington.

'You are changed since we sailed,' said Darwin, making a hesitant probe in return for Covington's probing him.

'Well, I have my reasons, I am glad to be back with 'um,' Covington jerked a thumb at his shipmates, who were preparing an upturned boat on the shingle and making a passable shelter for the night. But Darwin knew better—there being few secrets aboard when it came to men and their carnal temperature—and it was said his coarse Cobby had had him a whore and fallen the full-bottle. Darwin, being pricked with interest over the matter, was unable to make any straight reference, of course, otherwise he might crimson and splutter. What he *could* say was that he'd rather missed Covington in the way of a busy, bustling wife.

On their days of sleeping ashore Covington woke first, yawned, stretched and broke wind. He gave his blanket a whiplash to free it of burrs and dust before folding it and getting down on his knees to spark the tinderbox for an early fire. There was something irredeemably melancholy about a fire before sunrise, making a puny orange sputter against the fading stars. Then it became the best thing there could be. Darwin always said so and covered his head for his minute more, only to find, after that *momento* of fading dreams, his Bedford-born *vacciano* crouching over him, gripping a pannikin of tea and asking if he wanted shaving water.

'I can have it hot in a jiffy,' was the promise that Covington always made.

This was if they were near a river. (Otherwise they went dry, grew beards.) If they were near a town or coming to an estancia Covington searched around for a shining buckle where he could glimpse himself and apply his pomade, reeking of bay rum in the dewy morning like a Pall Mall rake. He always seemed to have recently broken his mirror and Darwin unwrapped his from its protective chamois and passed it over. Then Covington trimmed, with a pair of surgical scissors, his nascent moustachios, which were scrappy as a Chinaman's and yet enviable in the youth because worn with such expectant bravado. Darwin tried not to think about the woman Covington was believed to have tupped on his bird-shooting forays. He could do without the unsettling and unreliable effect of any passion that might break out in himself. But in the margin of a letter to his dear friend Fox he was caught sketching a seal of Cupid trimming the sails of a vessel. Covington found it when he readied the post, and grinned as if he'd found the product of a girl.

Gentry had to be pitied. They had so few advantages in respect of love. They could say they longed for a kiss from a bouncy wife in a vicarage garden. They couldn't say she roared under me and clutched my back, and I shot my specimen to blazes.

'*What are we about, you ask?*' The gent looked at Covington thoughtfully, not impatiently for once, and said they were about creeping back in time. He said that good spadework carried them there, and Covington, liking 'they', the way the gent used it so inclusively, boasted: 'Us Covingtons are bone-men going back a bit, you may swear on my pick.'

It was the day Gent wrote in his notebook with the writing slanting across his knee (Covington always looked, when the day's work was done, in case there was an estimation of him good or bad):

*My alteration in view of Geological nature of P. Alta is owing to more extended knowledge of country; it is principally instructive in showing that the bones necessarily were not coexistent with present shells, though old shells: they exist at M. Hermoso, pebbles from the beds of which occur in the gravel. Therefore such bones, if same as those at M. Hermoso, must be anterior to present shells. How much so,* Quien Sabe?

Nothing much there for a loyal retainer. Nor, as time went on, did Covington ever find his name written in the notes, though a *quien sabe* was something, having an echo of him, surely, and his perky Spanish. A speculation by the gent was sometimes followed with the phrase *vide specimens*, and as Covington was the source of the specimens it was acknowledgement of a kind. It was like a rock scratched with charcoal or engraved with a spike, the way sailors did wherever they went in the world. 'Cobby was here.' That kind of thing. He knew his time would come, as surely as he trusted in the completion of heaven.

Covington used his pickaxe to loosen a jawbone in the cliff-line. It was like a broken door. The creature had teeth knobblier than handles, suggesting a throat like a hallway and a stomach like a ballroom arch. It was turned to stone and embedded in soft rock. He took hold with all his strength and tugged remnants from the earth. Up came jawbone, thighbone, and at last the great skull of the monstrosity that so excited the gent he swore by his living bowels (which was most unlike him), declaring it to be a specimen of creation well off course. Indians they questioned reported no such animals ever roamed the spot, neither in memory nor legend. It was no African rhinoceros, either, as he'd thought at first.

Covington spaded towards nightfall, and then worked

the clunkering great nut free by lamplight and twirled the cranium above his head for all to admire until Gent raged at him be careful. It was to go to the Royal College of Surgeons in Lincoln's Inn Fields and be put on display, a rarity and a freak, and Covington was fearsomely proud of that.

The crew and even Capt smiled and made jokes of the cargoes of apparent rubbish brought aboard. But all of them knew that the taking of objects from earth, water and sky made a storehouse of treasure. The Spaniards had scoured the Americas for gold, whereas glory for England was in the naming. There were a few Frog Eaters gone ahead of them, one d'Orbigny was a particular foe, having had a three-year start. Gent was afraid d'Orbigny might get the cream of all good things. So they cracked on a pace. From birds to stones and bones and back to birds again, the mood was always the looking under of surfaces. Covington wasn't one to think with his head so much, that was established between them, but he could think with his hands and amaze with his production. Self-willed he might be as John Phipps had once hurtfully accused, but look what transpired, he came to this work and excelled. An easygoing servant without much pride was not for this master, and Covington meant to be prodigious and more. He would be made and shaped all over if that was the need.

The work ashore showed how far mighty golls, a solid trunk, and strong curiosity could take a young university man away from anyone's conception of him. Was Darwin truly to be a curate in a country parish, as was said? Those days with the monster-bones were a kind of bullock-hauling, man-driving display, with great organisation and demand on show. Darwin matched Covington in labour and there was no second-guessing: it was the most perfect of times and pray they came again.

'You are like my brothers,' Covington told him, adjusting the bandanna handkerchief tied above his eyes to keep the sweat from stinging. 'You moan and groan but you never tire.'

'I congratulate you,' rejoined Darwin, 'on being like your brothers too, I daresay. But I *am* tired enough, and will sit making my notes while you feed me my matter.'

*1st the Tarsi and Metatarsi very perfect of a Cavia: 2nd the upper jaw & head of some very large animal, with 4 square hollow molars & the head greatly produced in front. Enormous Armadillo? 3d the lower jaw of some large animal: 4th some large molar teeth. Enormous Rodentia? 5th also some smaller teeth belonging to the same order: &c &c.*

'Covington!'

'Sir?'

'As you gather the bones up care must be taken not to confuse the tallies. They are mingled with marine shells.'

'What say?'

'I said ... bones ... mingled ... '

'I had my eye on that,' said Covington, beginning to throw fresh shells away. 'You took them from the water. It is how you wanted them.'

'I do want them, can you hear me? But now they are mingled in the sacks.'

Covington stood over a sack. 'It will be an hour's work to separate them all.'

'I just want my friend to know, when they reach London, they are not the same ... '

'He would have to be blind ... '

'Very well. He would have to be blind. I shall make the point to him.'

Covington whistled and knew he was on the side of good sense. Darwin might say that presumption was the foe of intelligence, but a donkey was not allowed sixty pounds per annum including expenses only to *bray*.

*

They camped near the boneyard under their upturned boats listening to a ghostly wind and all wondering if any of the great animals that died there were still on the earth. Darwin brought out his books, Covington held the lamp for him, and he mumbled his suppositions of what the bones were.

Next day John Phipps came with the cutter and Covington took him to see the finds. They caught a snake on the way, by standing on its back. Its tail was terminated by a hard oval point that vibrated like a box of Lucifers. While Covington pinned it down Phipps clutched it behind the neck with his thumb and forefinger. Darwin thanked him and cut the snake open, and said it was equal in poison to the rattlesnake, and might have bitten him. Phipps laughed and said he put his trust in God. Phipps got excited about the bones, when all was explained, and said it was like doing business in great waters, or going down into the deep, was it not?—Covington winked at knowledge of this great text between them. 'Like being in the heart of the sea and going down to the bottoms of the mountains,' he replied. Covington was proud of Phipps as their bark's best countryman bar none—fowler of strong repute, immaculate poacher and something odd to have aboard, his fierce dissenting minister at large who would never own to his truest calling except under the stars, but treasured it inwardly while he worked at being coxswain and maintop-captain.

When Phipps was around the gent took care to explain himself with exceptional patience, valuing the antiquated, beard-stroking, nodding, analogy-loving man who slipped him questions. 'Why doth the fire fasten upon the candle-wick?' was a favourite to consider. 'Why doth the pelican pierce her own breast with her bill?'

Continuing their camp they talked into the next night too. You could hear the gent's brain cog-turning in the silence. He peopled the plain with horned and armour-plated creatures all snuffling along and, in the case of the giant armadillo, rupturing the earth with the force of an eighteen-pounder in its claws. It was easy to imagine the thunder of feet coming through gusts of wind, and when Covington asked, 'Was there men like them too? Say giants before there was us?' he got thoughtful silence for his reply. No giants *before*, was the true answer, for thus it was that the Bible said, that man was placed on the earth and all creatures assembled around him for his use. But giants *somewhere* was a possibility, for who knew all the mysteries of God's creation? They had not seen any in Tierra del Fuego but the natives spoke of them. Wasn't it what this voyage was for, to do some unravelling? When they sailed into the Pacific, which would be soon, what would they find there? It would be a long haul before they saw England again, they were all suddenly thinking, and to break the lonely mood MacCurdy's voice from the dark was heard snorting: 'There shall be giants again if *you* ever have young 'uns, Cobby, our man,' and even at such a hint Covington had no thought that he could have started a child with Mrs FitzGerald, nor remembered it was through a strange uncertainty of himself in relation to his gent that he had mingled into his passion a sense of rivalry, so that there was, if you fancied it, a kind of marriage afoot between one and the other.

John Phipps stayed the night ashore, arranging a folded coat for his pillow as so often in the past. Covington stretched beside him and Gent surprised them at their devotions, joining them at their Lord's prayer and bidding them good sleep.

Then they were all gathered back to the bark. She sailed, and Covington slipped away from a life on the land into a dimension of sea-living that wasn't like his old shipboard life at all—nor anything else he had known—for it was part comfort, part privilege, part a clerk's assiduous organisation. It was also part exile—his deafness making him an island. It was all summated in the words 'gentleman's gentleman', a strange vanity to have in those parts.

To Port Desire they sailed where the officers ransacked an Indian grave, looking for antiquarian remains. To Port St Julian, then, where they went out with the guns and shot salt-water-drinking guanacos and stumbled on an old brick-built Spanish oven. Nearby were the remains of a small wooden cross that was three hundred years old. Magellan had been there and executed some murderers, it was said, as also had Drake, doing his punishments and calling the island 'true justice'.

To the Straits of Magellan. To Port Famine, within a wet circle of latitude, where Covington sat in the poop cabin listening to rain crash down, making lists and waxing fat on good supplies. Plum jam, fresh bread, three spoons of sugar in his coffee and then a sneaked fourth. A good puff on his briarwood pipe whenever he wanted. A warmth in his trowser-region too when his mind ran over Mrs

FitzGerald, the details always violently present to his imagination. Out on the deck, faint cries of the survey-makers as they read depths, and the monkey-chatter disagreements of the midshipmen over their algebra. Faint was the word—only hearable if uttered close to Covington's ear; his left ear by preference, being slightly better than the right, being a little bit farther removed from exploding shot-powder.

On to Tierra del Fuego they sailed, and its maze of waterways, where they had previously sailed in wild storms, but now came into a calm.

'What a great useless animal a ship is,' said Darwin, 'without wind.' The sails hung heavy as wet washing.

'One sugar,' said Covington, mis-hearing absolutely. 'I had. And a half.'

They had a ration of it going, an absurd restriction. Covington always made out through confessing to a slight indulgence that stuffing himself excessively was quite beyond him. It was a servant's trick. On occasion it meant he could be blind steaming drunk while Darwin believed him abstemious to the limit of a single glass, perhaps two if pressed. He bent his head to his copying. *Entrance of creek, dark blue sandy clay much stratified dipping to NNW or N by W at about 6°.* So much on all this! The feeling of the past reaching out from behind and looming over the present like a shadowy wave. It was a restless deep inquiry they were on. There were times when it came with understanding and times when it came confused. There was no time when it came godless.

The roaming and adventuring in South America were coming to an end; there was only Chile left; and then the Galapagos Islands. *On the beach a succession of thin strata dipping at 15° to W by S—conglomerate quartz and jasper pebbles—with shells—vide specimens.* Covington's fingers scrabbled into the bowl and took another chunk of Jamaican molasses-brown that Capt's steward, Harry Fuller, had left for him. It was you scratch my back and I'll tickle yours between the two of them. Covington had so much sweetness in his coffee now that it made a syrup. He felt a bit sad. What was he mooning about? Mrs FitzGerald?

There was a comfortable mooning in that. He would like to be with her again but he did not wish it always. He would rather be where he was. *On the coast about 12 feet high, and in the conglom. teeth and thigh bone.* But then in a flush of anger he put pen to paper, and asked her—if she were free—would she be his wife?

There was a commotion on deck. It came to Covington through vibration and the long absence of others in the cabin. He went out to look. The rain had stopped. The surface of the sea was as silver as the back of a fish. Whales were passing and men were everywhere watching. It was not unusual but these were leaping from the water, every part of their body came into the air, and when they slapped down the tail made a noise like a ship's gun. Even Covington heard it. It was one of his last great sounds before all went inwards.

*Spermaceti*, he was to copy later. Also observations of the people who ran to the shore and paddled about in their flimsy canoes—their formerly civilised Fuegians who were absorbed back into savagery. *One full-aged woman bedaubed with white paint and quite naked, the rain and spray dripping from her body, their red skins filthy and greasy, their hair entangled, their voices discordant, their gesticulation violent and without any dignity.* What branch of creation did they come from? For what purpose were they created thus? *Viewing such men, one can hardly make oneself believe that they are fellow creatures placed in the same world.* So wrote Darwin, who was in Covington's care—Cobby with one eye open, his head resting on his arms, sentinel of the mess-table while an iron pot sang on the stove, always ready to spring alert when tapped on the shoulder (the bark being used to it now, when he didn't respond to shouting). There were days when nothing happened, when the gent 'staid on board,' when 'there was nothing to be done,' when the vessel 'remained stationary,' and the gent 'not being quite well' lay in a darkened bunk with Assam tea poultices on his eyeballs while Covington fetched steamed towels and potions for headaches and cramps. Problems poured into Darwin's brain and solutions withered at the rate of seeds, each one inscribed by the little curl of a question mark done in lampblack ink. Covington had charge of the notes and read them with all the understanding of a fly seeing a pinpoint of light in a dark room and dimly buzzing. *How to explain part of the structure of that Decapod? It is so very anomalous and the animal being pelagic is a beautiful structure for holding to light floating objects.* Quere *if a serpentine rock be not the produce of volcanic baking of a chloritic slate?*

There came further wonders surpassing imagination before their voyage was done—an earthquake in Chile dissolving the solid perfection of matter, turning forest and hillslope into substance reliable as water; God showing infinite humour, stranding seashells and a line of old beachfront high in the Andean mountains. What did they signify, these shells, if not this?—That the world rose high in perfect mockery of the Flood, as was philosophised between Gent and Capt?

John Phipps had a reply for 'em, if they might listen, an answer from which he never varied, that was good enough for brother Cobby: those beaches meant a playfulness in the mind of the Creator, a great teasing and tricksiness to test man's easy diversion from the Right Way. God rested on the Sabbath in perfect accord with the purpose of Creation, which was to attract praise for his deeds. And was there any contradiction to this in the way Darwin sent their specimens back by available ship to London? Not at all. The great ones of London turned them over in their hands and decided what they were—mirrors held to God's glory, mysteries of providence. The careful description and placement of material on lists connecting one to the other was elaborate praise. Collecting and praise arched together. They made the rainbow.

It was not for a naturalist to give so forthright an answer as Phipps did—though Darwin's reply on the matter was still godly enough: he said the Biblical calendar was not so accurate as one wished, and warranted considerable stretching to fit such foldings of the earth's coat to its ancient bones, that had, so interestingly, put a beach into the sky where mostly were rainbows. Things had lived on the earth much longer than anyone believed. God's patience was supreme. A-men.

On they had sailed to the weird Galapagos, the Encantadas or Enchanted Isles, so named because contrary currents bewitched shipmasters' intentions—the cold seas lapping the equator's burning hot sands, home of cactus, tortoise and lizard, where Covington shot Darwin's birds, and kept his own from two of the islands they visited, Chatham and James. It was the fourth year of their voyage by then. They marked the birds according to disparity in their beaks as mockingbirds, gross-beaks, fringilla, orioles, meadowlarks, blackbirds and wrens—though many were true fringilla or finches as was later made clear, which was the beginning of the end of Covington's calm, and the start of his shame.

It was a gentle breeze and a gloomy sky. It was an island lowly shaped, uniform except for sundry paps and hillocks, and when Covington and Darwin stepped ashore, there was small leafless brushwood to knock their knees and low trees offering no shade. Black rocks pulsating heat upwards. The sky hammering it back down. An unpleasant sappy smell in the vegetation. Insignificant, ugly little flowers more arctic than tropic. Lizards of the insinuating, crawling, belly-dragging kind, arousing hostility in the naturalist, who called them 'imps of darkness', and an impatience in Covington over getting and gutting one, because he was there for the birds. Small tame birds came to within three or four feet of them, and Midshipman King, in a separate boat party—still a raw boy and increasingly a stranger to Covington—threw a stone.

The stone whizzed past Covington's ear. He didn't see it—only the result. The birds were close together on the ground as a carpet sewn with lumps, and sheeny with the light of the vertical sun on their hundreds of backs. They had no predators and were so tame that the stone landed among them and Covington saw the dark birds ripple. None of them even flew up. Darwin prodded a hawk with his gun-barrel. Covington saw King toss his hat like a spinning disc and it landed on one. A mockingbird, it might be, though it could as easily prove to be a finch. King stood with his hands on his hips, laughing. A minute later he lifted the hat and found the bird dead. It was so easy to accomplish that he did it again. So did MacCracken years later and every visitor who came there. The birds were strangers to man and innocent, and so what was to be done with them? Covington spat with contempt. It was no place for a crafty birder.

But how many birds of the same kind did he take, just so? *All he could get.*

Birdskins were capital. Capital unemployed might be useless, but could never be worthless. Covington's imagination stretched like mastic and at the end of the join was London—Leadbeater's agency and a fine stack of guineas in a birdskin purse. So—what did the Galapagos offer him? Innumerable differently shaped beaks among birds of similar plumage! Rare takes! Common as dandelions in a spring field you may say! Well, all birds were common somewhere at some season and the point was, to Covington's brain, common captures were rare captures if you were the main one shooting. He went potting and dunting through a landscape studded with black cones and ancient volcanic chimneys formed from subterranean melted fluids. He shot multiples of birds from island to island as his master's needs dictated. The mockingbirds interested Darwin as differing between the islands, and these he instructed Covington to tag with their island names. The finches did not attract him this way. It did not seem important with the islands so close to each other, and the finches hard to tell apart. They were everywhere in the lowland thickets, milling around like flies. So Covington paid attention to the shipping tags of the birds he kept for his own, on which he marked, for his own sales' prospects, their island names. So did Harry Fuller, Capt's servant, mark what he shot, when Covington brought him into his game in a quiet way, and favoured him with a few directions on what he might back with his Capt's old gun.

They walked along in separate silences through truncated hillocks, black as the iron furnaces of Wolverhampton—a sea petrified in its worst moments, they likened it to—old craters a ring of cinders, subsided, collapsed. Darwin and Covington went on alone. Through these circular dips came the tortoises, creaking, shuffling, malodorous. They were billiard tables in extent and heavy as anchors. One was eating a cactus and quietly walked away. Another gave a deep loud hiss and drew back its head. Along came the boy-shooter now a man, and the man-boy naturalist—astounded, humbled, thrown back in time and both exactly startled in the same thought for almost the last time: that here they were surrounded by the black lava, the leafless shrubs and large cacti, facing the most old-fashioned antediluvian animals, or rather the inhabitants of some other planet, proof that God's hand sizzled here with one thing, there with another, and the chambers of his gallery were infinite in their on-going.

The *Beagle* was done with her surveying by the end of the Galapagos, but still had another year's sailing if you will believe it—Tahiti, the Bay of Islands, then across to Port Jackson, where they began their Australian visit, after which it would be the Indian Ocean, the Atlantic, the seas of Britain, the Scilly Light, England and home.

No angels announced themselves with hosannas when Covington first came to Sydney, only a long, low cloud of bushfire smoke, rotund as a fat cigar, streaming northward and scenting the wind. Approaching the coast, instead of beholding a verdant country as Covington had foolishly expected, he saw a line of rusty cliffs and a solitary light-house on a bare headland. An easy air carried them towards the entrance and around the point of the South Head. Thunder and lightning drove away the smoke but made the nerves uneasy. First impression was of a place covered by woods of thin scrubby trees that bespoke useless sterility. Approaching farther onwards, patches of country improved, and everywhere beautiful villas and white cottages were scattered along the beaches, and windmills stood on the ridges. Within Sydney Cove, where they anchored, there was a mood of incessant trade, and a great feeling of British industriousness to please the English eye. Darwin compared it with South America, where several hundred years of Spanish enterprise fell short of what the British had achieved here in several score years. But a damp was soon thrown over the whole scene by the news that there was not a single letter for the *Beagle*. Nothing for Covington and nought for anyone else either. They were like a ghost ship forgotten by the world and they had the world's secrets in their hold.

Darwin rode inland to see if he could find himself a platypus and to look at the geology. Covington found a pony and stalked the harbour headlands carrying his box and fly-net, gun, rock-hammer, and spirit bottles for reptiles. His three pairs of bird-scissors, as inseparable from him as his own fingers, were in a guanaco-leather pouch on his belt. Coming along the clifftops below the South Head light, he whistled and believed it might be a better place than he first thought. Money could be made with ease—the whole population being intent on acquiring wealth. A little capital could be trebled there in no time, the incessant topic of conversation being wool and sheep grazing. So while Covington swept the bushes (achieving, with almost every whack of his net, finds unique to science), he planted a few ideas in his head. The best was that servants soon became masters.

Then it was the home run after their time away—holds laden with pressed twigs and petrified wood, stuffed mammals, corals, reptiles and bottled fish. On the deck a live tortoise and Phipps's Falkland Islands fox. Delays made the hungry soul more fervent. Patience applied to work and to God, but hardly applied to getting home to England. It could not come too soon for any of them.

'If a man would live well,' preached their Capt, 'let him fetch his last day to him, and make it always his company-keeper. He that forgets a friend is ungrateful unto him; but he that forgets his Saviour is unmerciful to himself. He that lives in sin, and looks for happiness hereafter, is like him that soweth cockle, and thinks to fill his barn with wheat or barley.'

Gratitude was unbounded; the practice of gratitude expanded faith, right to this last leg; in faith and fellowship they moved about the bark on those scented, starry nights as they wallowed up through the tropic Atlantic, heads full of reunions with loved ones and English beer that would be theirs to guzzle in a few short weeks, and thoughts of maids to gentle—which did not exclude their Capt, who had a fiancée waiting for him but had never mentioned her once in their full five years afloat.

One day a gun went off in Covington's face and burned him, even though there was never any foolishness with weapons in his style of work. He had observed all the rules and never put a gun down loaded, or cocked, never pointed a loaded gun at anything unless he meant to fire, and never climbed through rigging with a loaded gun either. So he ensured he would never be taken by surprise.

The piece that he loosed at a gull that day ripped apart, throwing fragments of hot steel back over his scalp, and setting the skin of his right cheek alight with searing powder. The surgeon smeared his face with linseed mash and put his jaw in a sling.

In the poop cabin their work was barely disturbed. 'I have had worse flayings, mark you,' mumbled Covington in defiance. It was time for organising papers and sorting ideas on what the voyage meant. The question of Creation was strong in the air, for their cargo was the cream of it. They had a kind of vanity around the subject—having sailed farther, longer, and been more punctiliously obser-vant than any great voyage, and done it in so gentlemanly and offhand a fashion, considering what they took into their hull. It was time for Covington to boast his copper-plate once more and wonder to himself if Mrs FitzGerald had ever made her way to England after he'd made his demand in the stormy letter posted from Chile. He rather believed she had not received that squib, as there was no mail waiting when they reached Sydney. She'd be an old woman by now—at least twenty-nine.

Coming up the Atlantic they stopped at St Helena and he learned his Hickory maid was dead. A burning pox had taken her. The glade where they had lain together was overrun with blackberry bushes. Although he shed a tear, so much of Covington's past had happened in another age that it seemed his rough black sweetheart must have lived a full life, and a kind, if not such a good one since the last time he had shipped through. Because all had their time, all passed from their time. It was a rule true of the bones of ancient beasts as it was of men. God gave, and the species of life had their day. At twenty years of age Syms Covington felt one hundred years gone from England. When he spouted his *Pilgrim's Progress* with Phipps it was like old John Bunyan was someone he knew as a boy. Parted away from himself he acquired a thick, defensive shyness. He wore a path around the vessel; loomed on people day or night. Main hatchway, fore hatchway, measure the spirits of wine, be sure none was filched, be sure the packing papers were dry, check the powder store, oil the guns, check for mould, seek rising damp, air the birds, make more space, think ahead. Was it just the vanity of rubbing shoulders with professors that made him sharper of mind? He was on the track of a suspicion. His outer parts were distracted, staring, vacant, inattentive except to his gent's needs. Inwardly he burned. Thus did 'What is life?' become a question in the mind of an ordinary young man—Syms Covington of the *Beagle*, who perhaps from joy in life should not have been asking such questions at all, and whoever planted the seed of dismay in his heart should have been tried for a crime.

Darwin went to Sunday service and bellowed all the hymns, sometimes with tears in his eyes. But when he was by himself he was like the robin in the parable who gobbled up spiders, drank iniquity and swallowed down sin like water. Which is to say he thought of other matters, changed his mind plenty, and so did his servant through matching his needs get a good hint of it.

Soon it would be time to give a report and Darwin wasn't sure he could give it without his expert friends in England making judgements on the collection. The mystery of mysteries was in determining how the world worked. How creation *did*. They would all have a hand in it, his Lyells, Henslows, Sedgwicks, Hookers and Goulds. With a great principle at stake, all the dryness of minute specific comparisons vanished in such men's eyes. They either believed, or they did not. But on which side?

Covington picked at his scabby cheek and wondered.

He was nothing to the gentry, but all the same much came to him from the unseen. He didn't need to have things explained to him out loud any more. There was plenty in his master's notebooks of listing a fact and then probing and questioning around it. Deafness made a difference—positively, you might say. Covington, slow as he knew himself, moved in a different medium from the one he had inhabited with his hearing intact. Their voyage had a meaning that crept over him without his wanting it. It was like the switching of tide in the darkness while standing hip-deep in a dark estuary: how much more patently the current could be felt then, than in the distracting glare of day.

He knew this: great questions were getting asked in Darwin's head. *How did it start? Where did their booty all come from? You could say God, but if that was the case, and God made us animal by animal and saw we were good, why did some die off? Why were extinct creatures the relatives of living ones? Why hadn't God just started a clean slate each time?*

Covington had a good raw taste of it when he copied the

zoological notes in those last weeks coming home. The poop cabin was a welter of papers and the two of them barely saw daylight except for going outside to stretch and sniff for land. Darwin made a note of the varieties of tortoises and birds from the Galapagos and how they radiated out from their centres of Creation into the several islands. The mockingbirds they had collected were singular from existing as distinct species in the different islands. They were allied to the Callandra of Buenos Ayres but set Darwin's brain buzzing with the thought of how each one was constant in its own island. He puzzled over the mess his brain was in, and wasn't blaming his servant so much, but would rather have liked to. They might have looked closer into those crowds of finches and likewise the tortoises, might they not?

The governor of the Galapagos had told Darwin that Spaniards were able to tell which island a tortoise was from just by looking at its shell. Darwin thought at the time it was an interesting comment on varieties, but now he thought, *What if they had been species instead?* What would that say? But it was too late. When they left the Galapagos they had taken a good number of live tortoises aboard and ate them as they crossed the Pacific.

Darwin addressed Covington through the hard plates of air that constructed deafness around a person:

*What did I say to you about them, the tortoises?*

'You said, "Anything saved?"'

*Why, didn't I ask you to note what islands they were from, and have you make sketches of their shells, and dry and keep a few carapaces?*

'It was too late,' snorted Covington, 'if so I would have done it—they's been tipped overboard after the steaks was gobbled and the soups all supped.'

The same thing went for the finches that Covington had bagged for his master—hardly a one marked by island— and why should they have been? Jehovah himself would assume that the islands were so close together that no reason was possible for them harbouring different species. Although: *If they weren't just varieties* (Darwin jotted, his mind circling back on him) *then such facts undermine the stability of species.*

Covington was quick enough to ask if that meant a certain word. He'd heard Darwin say that transmutation was the catchcry of doubters.

Darwin looked up from his papers. Covington repeated his question, fingers stained with ink that he wiped with a twist of rag. *Transmutation?*

'Is that not the meaning of it, to *undermine?*'

Darwin stared back at him straight, in one of those looks he threw from time to time, that had mighty cog-wheels turning under them. He spared Covington little in the complications of his mind because what would Covington understand anyway? Besides, the shooter was conveniently deaf. And thirdly, if you must know, was a servant, and if a man could not be himself with a servant then he would lead a life of arch pretence.

*You know very well. To undermine is to make a gift to doubters. Remember the bone hunt of Punta Alta? It framed a similar puzzle, because if species can change— evolve—transmute—replace each other—where does that leave Creation? Either God's handiwork is perfect through time, or it is not God's handiwork at all.*

Covington felt a coldness in his stomach like a lump of suet pudding.

Darwin leaned forward over Covington's shoulder and made a nice addition to his jottings, adding the word *would* between 'facts' and 'undermine', and giving Covington a faint smile as if to say, *Satisfied?* So that now it read: *If they weren't just varieties then such facts would undermine the stability of species.*

So the upshot was good in Covington's mind: the master was a Christian young man, to be certain. 'Would' did not mean 'did'. Covington, who lived with him close as a wife, concluded so. John Phipps concluded so, too, and had his proof in a conversation they had about what could be shifted from one part of the world to the next by animals, apart from barnacles on the bottom of a ship and fleas on a skinny Falklands fox. They recalled a Captain Henry they'd met in Tahiti, whose father was a missionary. He was a curious young man with a fund of observations. His greatest disdain was for American whalers from New Bedford. He said they were full of watery Congregationalists, an opinion that made Phipps and Covington smile, for that was their own denomination when they had cause to declare what it was.

Capt Henry said that a shark had followed him from the Friendly Isles to Port Jackson, that he'd caught it inside the Heads, and lo, there were things inside it that had been pitched overboard early on—a ladies' purse made from purple silk clasped with iron, a ball of string, a crockery plate marked with the crest of the Henrys of East Anglia. Darwin said, 'I must look in a shark more keenly.' He wrote in his notebook that Dampier in New Holland opened a shark and found the head and bones of a hippopotamus, its maw was like a leather sack, so very thick and tough that a sharp knife could not cut it. The hairy lips were still sound and not petrified. The maw was full of jelly that stank extremely. She was hooked at Shark's Bay, latitude twenty-five degrees. The nearest of the East Indian Islands, namely Java, was one thousand miles distant. So Darwin asked: 'Where are hippopotami found in that archipelago? Such have never been observed in Australia. So there it is, with respect to sharks distributing fossil remains, I think it is likely they do, Phipps, soft tissue or hard plates, for how else could the tragedy of the Flood be spread so wide?'

C ovington stepped onto English soil a mauled, weather-battered survivor. Darwin was not so scratched about, but brown, healthy, excited and full of plans—a bachelor of twenty-eight notable in his particular circles, and most secretly anxious to mate with a female, as Covington happened to know; also quick to enlarge his advantages in the world of natural philosophy by spreading about what they had found. Over his years away friends had read his letters to learned gatherings and extracts from his notes had influenced the great. He was their pair of eyes and a brain that was attached to a mainmast and sent scudding around the world while others of his class did it rather more gently. He left the *Beagle* at Falmouth quivering with fame, while Covington packed the specimens securely, got drunk first chance he could, and returned to the vessel to see that nobody took or spoiled what they had until it could be properly unloaded in the Thames.

It gave Covington a sight of the Celestial City when they sailed up the river a fortnight later. Bedford was only fifty miles away and he'd been round the world to get there, yet had never been to London, not once in his life. It showed the difference between being told about somewhere and seeing it with your own eyes. London was all covered in grime. Instead of shining, the city smoked. Soot floated down like black snowflakes. Iron-rimmed cartwheels clattered through the streets and Covington felt the vibrations of them all around him, there was no need of warning shouts. Instead of gorgeous raiment the people wore dull cloth. There were beggars in every doorway; bold thieves; outright pickpockets. Those on day wages were the lucky ones, while the poor milled through, glowered, starved and drank themselves insensible when they got the chance.

Darwin asked Covington to continue in personal service and he agreed most gladly. He started the day fetching shaving water and ended it bearing a hot brick wrapped in flannel to his master's bed. Was he to be merely a footman? His employment being of such a mixed nature you would think so. When asked his station he puffed a bit, saying he was a gentleman's clerk or amanuensis. 'I was with Mr Darwin on the *Beagle*, and he relied on me somewhat ...' Darwin's friends, the experts of London nature study—Lyell, Owen, Gould—referred to him as the Trusted Cobby, the Right-Hand Man or Hefter. They treated him as simple and went on with joshing. It was well meant between each other, but didn't fit if you weren't their equal. Darwin's brother Erasmus always greeted him as the Good Shooter Covingtonius, and asked him if he kept his powder dry, and then laughed and clapped him on the back. Before Christmas master and servant took leave of each other and went their separate ways to see their families.

Covington went part of the way by coach, and then by foot. It was cold and damp as he came by the flatlands near Elstow, a sack over his shoulder, a stone in his shoe. Closer to home his heart sank. The hiss of footsteps across dewy-damp grass. Empty houses. Graveyards of names. His small nephews and nieces dead of fevers while he thrived in foreign climates, their only luck that they knew heaven first. A door-hinge creaked as he entered the old chapel, peering, his eyes lifting, discovering that the window he loved, high in a wall, was gone—and some greatness in his heart leaping the obstacles of the world was gone also. The gap was nailed with boards. Bedford and what Covington loved there were displaced. He came to the other end of Mill Lane, the door of the small house, the gathering of his closest. His brothers were short of work. His Pa was come on hard times. Mrs H was so sickly as to barely smile at his kiss. Yet she basked in his gaze as if he were sunshine. There was so much to tell, but where to start? He made boasts of the higher-ups whose serving class he had attained. His Darwin came over somewhat sharp, his FitzRoy as grand. He told them stories of Sydney Town where a man was as good as his purse, and heard his voice get excitement, and they were all at him at once to emigrate and send for them when he was settled. For the rest, the magnitude of their chase, he was dumb.

There was a letter waiting for him from Mrs FitzGerald. His fingers trembled tearing it open. She told him her news. She had a daughter. Theodora, beloved of God. And so Covington held in his arms, when he prayed, the sense of a white, penetrating light, a cradle of light, a basket! He filled his arms with this light as though it were silk. It streamed into him redeeming him from water, from gunshot, from doubt, from the damage to the nerves of his face. *I am very fortunate,* Mrs FitzGerald had written. *She is a lovely, clever child. Should it ever move to you to send a remembrance, do it through Mr Lumb of Buenos Ayres.* She signed herself his cousin, which he took as his cue, sending a child's bonnet and a money order for five pounds, which was a great fortune for him to spare, and signing himself 'your cousin' in reply. When I am able I will send for her, Covington resolved. He never once swayed from the resolution, either, and at long last Theodora was to come: and it was proved, she was his, and this light of the world he held in his arms proved eternal.

It was always his plan to go to the old chestnut tree, where he and John Phipps had prayed, and meet there with his old shipmate. But Phipps took off in another ship, they missed their goodbye. Where Phipps went Covington only heard by rumour, until the last day came and they were met again in a strange fashion, and Phipps preached a brief sermon, restoring Covington's soul and redeeming him to all of creation.

After Bedford he joined Darwin in Cambridge, unpacking quantities of specimens and sending them around. In the New Year Darwin took a house in London. Covington worked in the basement and slept in the attic.

There was a second accident:

It was at the Royal College of Surgeons in Lincoln's Inn Fields. It was a veritable mortuary of bones in there. A block and tackle with a broken link came loose and the crate they were winching crashed from an upper door. It made a wild ringing as it smashed against the walls, flying from side to side on coming down. Heads appeared at windows and there were shouts of warning, but do you think Covington heard them? Not he with his deaf man's look of waiting to be surprised.

The middle fingers of his right hand were crushed. They were stiff ever after, and when he raised his right arm to comb his hair or trim his whiskers the tendons in his right hand, somewhat damaged, caused his index finger to lift back, so that he seemed to make a beckoning gesture to himself. 'Follow me.'

Well, it was sad to think of it, that it might come around to a misery of doubt and anxiety regarding God; that a man might have only himself on this earth as a guide whatever his heart told him, and that such a loneliness might be proved, and that the man to prove it had a servant, an accomplice in the affair, and his name was Covington.

There was much in this London time that Covington wasn't meant to know. And much that passed him by. But how could he not know certain matters after his footman's day—when he straightened a page at his elbow, spread a fresh sheet of parchment in front of him, and began reading Darwin's scribbles and transferring them across in awkward copperplate? He was back to listing hides from South America as in his boyhood, only now it was bird-hides and mouse-hides. While the work elevated him in the world somewhat, it also put him down. Tucking his tongue in the corner of his mouth, twisting his ankles around the stem of a stool, he gave all his concentrativeness to the task. Darwin often checked over his shoulder. A sooty cat looked in from the other side of the window. Coal smouldered in the grate with sickly bloodshot insufficiency. Covington fell asleep over what he was doing and woke with a start, finding his pen still held in his damaged claw. Darwin sat in the corner with his head thrown back, his fists clenched, doing his thinking.

The lists of what had been collected on their voyage were prodigious. It was taken by Covington, shot by him, smothered, clubbed, spiked, dug. It was skinned by him, boned, plucked, pickled, dried. Afterwards packed, wrapped, labelled, stacked. Then hefted, carried, carted, deposited and shipped. Finally unpacked, re-sorted, transferred, and

brought to light. Men came around for the sorting, expressing interest. Strange that once he'd thought the collecting the
whole meaning of what he did. And that was only his first
vanity. The material had been sent back to England for the
glory of Britain and the interest of science, and he had never
decided what the interest of science meant, except that bones
were measured and fitted to other bones until declared to be
megatheriums or macrauchenias or whatever. In this reckoning of booty everything went Darwin's way, and that is how
it was between master and servant. Then it came trembling
up against questions, touching the one inequality between
them that tipped back in Covington's direction—his four
finches labelled by island whereas Darwin's were not.

It was March by then, five months after their return.
Darwin snapped his eyes and asked Covington if he remembered something. *Had* he labelled his birds by island?

Repeating the question and waving his arms.

'Covington—your birds, if you will.'

'What birds are those, sir?'

'The finches you took on the islands. Are they sold yet?'

'What islands?'

'You know the islands I mean, Covington. The
Galapagos chain. Out of the hundreds you took on your
own account.'

'You weren't to know of it,' Covington thought to
himself.

'There were no secrets on a bark of six guns,' Darwin
implied in his softest smile.

'You never said not to turn a small profit,' Covington
purred to his inner conscience.

'Gentlemen don't always,' he imagined Darwin thinking.
'They have their restraint. They just mark their servants
down.'

'You were not so interested in the finches there,'
Covington said brazenly. 'You just sent me around blasting,
and I did your will. You wrote in your diary, which I have

just made in fair copy, that it would be interesting to find from future comparison what district or centre of creation the organised beings of that archipelago must be attached. You never thought the different islands mattered because a godly man would only say the creatures were right and proper for where they lodged.'

'Why the cavilling, man? Are you becoming a pedant, my schoolmaster? I never knew you to take such an interest! Now tell me—what say your labels?'

'You doubt Creation.'

'Still your thoughts.'

'Is this on God?'

'It is on the geography of finches.'

'Charles Island,' Covington said at last. 'And Chatham.'

'Ah. Very good. Very good. And Harry Fuller's were shot on Chatham and James. Bring yours to me, then.'

'But they are mine.'

'Whose employ are you in?'

He brought them down from his attic room: a prize four of finches neatly tagged, their island homes noted and dated on the tag as Beskey had advised, to give cachet to his find. Such lumpy little birds they were, about as interesting as starlings, the female speckled somewhat, the male rather dark.

So to the dim drawing room on a day soon after, when Covington stood obediently, sullenly holding a scrap of paper in his hand, barely able to read Darwin's lips in such light. Windows dark, wet with yellow fog, lamps lit throughout the day, coalfires helpless against the bronchial chill. The receipt was for Covington's four finches delivered to John Gould at the Zoological Society, and yes, taken round to him loyally. Covington believed he was to have them back again, his *Geospiza magnirostris* as fine a demonstration of the taxidermist's art as any collector would know. They were Covington's own sworn property, portion of his small estate, dull in plumage but valued by Leadbeater's at several guineas apiece. No matter, they were gone from him, taken to be used as proof—if you please— of the order of nature. A most crucial proof as he learned from Darwin's wanting the specimens so badly, though whether it was to be a fully reliable proof he was afraid to know.

There was anger in the dull house. The matter was to stay secret.

Darwin said it wasn't *his* doing, if it was shown the Bible wasn't true. It was the nature of beings and their stations in life that would speak the blasphemies if there were any. Darwin wasn't in the business of proving atheism, he said, or anything else for that matter that would undermine Creation, but was only setting his mind to the material, in much the same way Covington cupped a hand to his ear— to hear better, was it not, to hear *exactly* what was said?

Then the whole of Covington's life seemed merely a story he told himself. Why should his outdated existence be exempt from condemnation and saved for God?

There was the light on the face of the waters. There were the seven days of creation. There was Bedford, the slow river and the lock gates dripping water. There was his captivating by Phipps, and their eager voyaging deep as a dream and lost to his waking mind. Then the *Beagle*, the full seeking of service with Darwin, and the never taking no for an answer until he was taken on.

And the next part of his life, too. Also a story he told himself. Staying with Darwin another two years. Getting his golden guinea when Darwin, without warning, announced he was to marry and wouldn't be needing a 'wife' any more to warm his head plasters and mend his slippers. Getting his recommendations as an emigrant at the age of twenty-three. It was something to have been found 'perfectly satisfactory' at the end, 'generally useful', 'prudent', 'economical', and 'never once seen in the least degree affected by any spirituous liquor and trustworthy in the highest degree'.

In short to have been Mr Darwin's most obedient servant.

BOOK

5

*On a Journey South*

# 1 May 1860

It was almost morning of the next day. It seemed much longer since the boy drowned and MacCracken became their guest. Mrs Covington rolled from bed, wrapped herself in a blanket and went to the kitchen and kindled the fire. She stood warming her hands to the flames and considered the day ahead. It would not be an easy time of it, although what day ever was with Mr Covington as one's chosen in life? It would be into Sydney Town with all of them this morning, securing horses and loading the spring cart and pony trap. Good riddance to Sydney, then, by afternoon, and so making their way over badly made roads they would go, to endure who knew what frights until they came late to an inn. That was if what Mr Covington had in mind came to pass, and it usually did, and so there would be little comfort in their wayside halts except for sacking beds and smoky kitchens and salt beef and sooty bread on the table. Days of it to follow as they wended their way south. Dust like talcum in the nose and ears. Flies to be swallowed and endured. And if it rained, a quagmire of bogs and rivers to be crossed. Mrs Covington was raised to such things, being colonial-born. She bore them stoically enough. But she would rather they travelled by schooner, and in a stateroom, too. Then there would be just the sailing, the disembarkation at the wharves, the short foray inland and the steep haul up ridges by bullock-cart, with bellbirds in gullies and the shade of tree-ferns at morning-tea stopping places. A return to the country was what Mrs Covington wanted—had argued for with her husband until she had no words left. But only the arriving—that was what she cherished—never the getting there. It was a relief that now even he could see it was time to bring his whims over 'Coral Sands' to a finish, because they had a home, and it was in the wilds, and she longed for it.

They had not been back in almost a year. Mrs Covington thought about what it meant to her. When she arrived at the top of the ridge where the farm buildings clustered she would shake out her rugs and call up her pets, and introduce Theodora to her cockatoo, who called everyone by name. Her piglets would be thick-shouldered boars by now, her poddy calves milkers, her housemaids would all be run back to the wild bush and be about bearing children to bearded Irishmen. Her sons—what word of them ever came, except as relating to cattle and land?—they went chasing to the edge of the world and made her terrified with thoughts of the dangers they faced, leaping their horses over ravines and plunging down cliffs. When far from home they slept with backs to the fire and their carbines loaded in defence from creeping savages. Her daughters—what about *them*? Untameable freckled whipsticks competing with the boys, stubborn to be left with the older ones when she tried to get them to the city and the pretensions of comfortable life. She could hardly blame them.

How she loved the clear light on the ridges and the sound of the river below, and the rock where she sometimes took herself, perching over the drop allowing cool breezes to fan upwards through her dimpled knees. How she loved the simple joys of steering a cow back to pasture with a springy sapling, the churning of butter and the sour damp wood of the dairy, the finding of warm brown eggs and getting them in her basket before goannas found and broke them, licking them empty. But she was bound to her husband by her vows, and her love, and if their bush-bred brood were repelled by his clumsy hopes, his Covingtonisms, then stand by him she must.

She *hoped* Theodora would come. She would introduce her, with a blush, as her daughter. If neighbours smirked over that she would redden even more, no doubt, and be uncomfortable with the truth, but hold their eyes until they thought better of their snickering. Theodora might teach her half-brothers a few manners, too, if they ever came in from their roaring lives. What Theodora might do for the daughters, advancing their manners, was too wishful to be imagined. As for Dr MacCracken, who Covington was hopeful to bring, he was the reason they were to take the overland route, so that he could eye the prospects for land. Although he considered Mrs Covington an old eccentricity, she believed, and barely worth two hellos of a morning, she desperately hoped he would come anyway, trussed up like a cornstalk or no, because, *Thank God for him*, she thought: *I have a husband who's as mad as May-butter. But I dearly love him and would have him rescued for his soul, if such a thing be possible in a man.*

First greying of morning. Scattering of dreams across the four seas. In the main bedroom of 'Coral Sands' Mr Covington rolled to the wall, pressed his hands prayerlike under his chin, and groaned that he would never get back to sleep at this rate. The events of the drowning hung over him like a swarm of bees. They had entered his dreams. Which boy was it, then, that got stung to death in a watery hive? Darwin's book was in the swarm like a sheet of wax comb. It announced itself in a waking apprehension of the guts—hell's honey-maker, and he avoided thinking about it as best he could, half-asleep, holding the matter at bay. Darwin. Let him go staying down in purgatory pumping thunder, as was said in ships. No wonder Covington's mind baulked, ducked, and ran away—yet was all the time drawing closer to the heart of the question: *Whatever shall appear on fair enquiry shall be the truth for there cannot be two truths.* Nothing had changed about the question for years, except this drawing ever-closer to a full answer.

Covington thought: to be clubbed over the head and made stupid would be a right way for me to escape this pass. *So pray make it swift.* A single blow as to a beast—he could show an accomplice the exact point on the skull to bring it about, placing his finger there to mark the suture. Should he call MacCracken, and get him to act from pity? That was how low he was. He sourly thought: *A man is only a beast, is he not? Only rubbish in a boneyard. If it be so proved.* Then, almost as quick as Covington dozed and thought of oblivion, he was thick in his armadillo dream.

It was over-familiar to him.

It started with an ostrich and he was on horseback in South America. Standing in the stirrups he made a trumpeting noise, scaring the bird from its nest in the grass. It stretched its wings hawk-like and half ran, half flew with the wind catching underneath and its legs skimming across the ground until it soared and disappeared into the sun. The *Rhea darwinii.* God help him, he thought, coming awake a little, perchance it wouldn't be the armadillo this time. Perchance he would just go up this green rise in front of him, kick grass from his boots like a trapdoor shut, and so enter heaven. When he went back to the ostrich nest to count the eggs he saw a striped butterfly flitting in the grass, and pursued it on his knees. When that was despatched to a pill box—with a pin through its thorax—he picked the flower that went with it. So was he free? No, because lifting his head he came upon the animal—a glowering, humbled, armorial beast that filled him with fear. It was easy to grab, but as soon as he tugged the hard tail the animal dug down, spitting damp soil from under its leathery skirts and rocking slightly from side to side. Covington sat with his legs apart and arms straining. This part was always a great tussle and he always lost. The creature was powerful under his hands. It roused his blood and sometimes from the dream he woke full proud and desirous. But not this time. Mrs Covington would be spared his manly grunts and the explorations of his hands this time. Now there was just disturbed soil somewhat quaking at the centre as the creature rocked to deeper safety. Covington rolled on the mound. It was how the wheel of a sinking ship must feel pulling a man down, except instead of the smooth flow of water there was smothering dirt. Crumbs itched him. A patch of armour-plating showed mockingly in the earth. The last part of the armadillo was small as a tobacco tin and getting smaller. He found that with his fingers under the shell the creature began pulling him under. *It cannot be happening* was his common response—his arms stretching and straining. At the end, all that showed of the animal's hard case was a piece the size of a button.

He roared awake:

'Me boots!'

And his wife was at his bedside with his boots warmed from the stove. He put them on, laced them tight, and sat looking down between his knees. It was done before he realised the boots would have to come off again, because he was still in his night-shirt, trembling and sweating. She returned, the old darlin', with his pannikin of tea and touched the back of his neck. *You're awake, it were just a dream.* He raised his eyes and thanked her. 'It was that infernal diggin' animal,' he said. He was wearing a red nightcap and his eyes were doleful. She had served the same loving function to their children and it seemed he was merely one of them in this state. Her kindliness wrung him through and cheered his heart.

C ovington went in to see his guest. MacCracken was fast asleep, his bruised cheek cupped in the palm of his hand. *Safe* was how Covington thought of him, the same feeling—if he cared to think of it—that he had around his prize bullocks and other breeding livestock when he had them yarded. During the night MacCracken had used the chamber pot, half-missing the bowl. A stain almost reached Darwin's book, which Covington picked up ready to throw away, but as soon as he touched it felt differently. The book had a changed threat after a black night. A hot confused pride spoke to his feelings. Half of Covington was in its pages—those years when his life was disgorged at Darwin's feet—the bundles of bones and birdskins to be interpreted—the glass jars devotedly sealed—the million gouged eyes—the innumerable notes copied in his own true hand. In this he was like a child peering in shadows, daring himself on. He tucked the *Origin* under his arm. He reached down and gently pulled the bedcovers around MacCracken's shoulders. Let him sleep until they were ready to pack, he resolved, and then surprise him by taking him along. He'd often wanted to come. Now was his time.

Covington went back to his room and sat on the edge of the bed. He turned the *Origin* over in his hands, and then pushed it under the bed. It was no good upsetting Mrs

Covington any more by letting her see him pawing it over this way. Just by bringing the book away from MacCracken boosted his courage. Well, he never *was* much of a coward, to tell himself the truth. Last night when he burst from the house and kicked the door shut behind him the vibration around the doorframe registered with his deafness, he'd whacked it so hard. He'd felt disgraceful. It must have despaired them all. He'd set off, taking a deep, humid suck of air into his lungs. Let no-one follow him, he'd thought, or they'd get a taste of his malice. The missus and Nurse Parkington were ever-pursuing him, coming up without warning, plucking him by the shoulders. He didn't want anyone to see him like that, enraged and submissive, thunderously appalled, helplessly wordless—trapped in confusion and to his great humiliation weeping copious salt tears. MacCracken had already caught an eyeful of him. It was just preposterous. He loosened the cravat from around his neck and mopped his cheeks. What could he do? Darwin's book had come and the boy he'd pulled from the water that morning was dead. Grief for a child was the warrant of his pain. Where did they go, those playful spirits? He carried a clutch of them round in his heart all these years. Waiting for the signal flag from heaven, he supposed they were. Joey Middleton, his sailor-boy friend, he was the longest—until his rags had seemed to rot from him, dear God, and what was Joey now? Would he still be bones? Would the bones be petrified and laid down in strata? When he saw the boy Pickastick dead on the sand he saw Joey again. God help him—when he'd clipped MacCracken's jaw he struck all those who held him back from diving in that first time, even counting old John Phipps.

The damnable truth was that his emotions always felt new to him. Here they were every morning sitting in his chest freshly charged. The world wasn't old at all while it had Mr Covington in it. And this was the worst time of it he'd ever had without doubt—since the letter that said the book of creation was almost done. That it was on its way soon. Then, God dammit, that it was in his house.

The dark long light of the harbour had been made of woodsmoke, sandstone and heath. He'd wandered around aimlessly and sat on his rock. *Smite* the view. He hadn't been there to witness any evening changes and appreciate them at all. What a lugubrious aching misfit he felt, an enemy of nature. He was there to get his strength up, to shake himself into fitness before going back in and using MacCracken as an arbiter of understanding. He didn't like those twittery Afghans but he trusted MacCracken. He remembered his first sight of him, from upside down when he was carried to the cottage on a plank. MacCracken raised his fingers and beckoned the sailors up the path: 'Quick now, and I'll save him.' Had Covington heard the bull of Elstow bellowing in MacCracken's rocky pasture he would not have been surprised. That was the state of mind he was in. Ready to go.

But he hadn't gone. He'd touched the walls of Mac-Cracken's sickroom, he'd followed the play of shadows, he'd sniffed the gritty aroma of sandstone country. He knew it back through his bones from the first time he came there. Sydney heathlands. Vegetation reeking of aromatic oils in the crackling heat of summer. When he started taking walks in his weeks of recovery it scratched, pricked, lurked. Its species were prolific and quaint. They lured his nostrils with subtle odours at every turn. He beat the bushes with his bug net venting the pride of the undefeated. Already the decision was made to spill his story to MacCracken. Indeed, he thought, he *had* begun letting it out in as many words, but this wasn't how MacCracken heard it at all, the times he showed himself on the hill, wild to collect anything and retaining every small boy in the east of Sydney to bring him their finds. Covington had jars of marsupial foetuses preserved in spirits of wine in his pocket some days, and MacCracken didn't know it. They were the tiny born young who lived on the outside of mother bandicoots, wombats and roos. But he couldn't bring them out and say what they were. The image of angels, so pink, folded and heart-breaking? He was mad enough to fancy so. It was easier to cover MacCracken with blessings and draw him into his wealth of timber dealings and land, expressing his gratitude for having saved his life, rather than coming out plain with what he wanted:

Which was to make his confession—that he was once an accomplice to a great murder. That the murder may not have yet been entirely done, and he lacked the wit to know how proved it was. There were no words in Covington's lexicon to pin himself down. He was a man of action—yet strangled by philosophy you might say. And that was a strangling indeed. Everything was mixed and contradictory in Covington's mind. Darwin wrote asking for specimens and Covington, always knowing what was afoot in the man's thinking, sent him whatever he asked. There was pride in that. Just as there was pride in being wealthy and in having the same number of children as Darwin (and one better, through counting who was born on the wrong side of the sheets). Covington was the prodigious collector after all. Yet he was in desperate longing for sending a great surprise—the winking eye of God, you may call it—all those bulgy swollen pink marsupial eyes a case in point, peering out of their glass preservatives. And so he sent what Darwin didn't ask for, too. Bombarded him with life, you might say, as he had done when they first started collecting together.

Always so full of *hope*.


Theodora came and found him. So dark it was and she carried no lamp. She took his hand and didn't get his attention at all. He was all in his head, and so she waited. She was at his side, but didn't know where he was. Well, he was lonely and sitting under an old chestnut tree in winter light, and making a prayer for company and good cheer. He was more spirit then, he mused, singing in his bones, a follower of heroes and at ease with animal nature, which he never doubted as a gift of God. He sighed for the immensity of a loss that he was even now unable to concede. (The sigh sounded to Theodora like a gust of wind, and she shuddered—what ugly old man was this?—but he still didn't notice her.) There wasn't a particle of him back in those early times that was given to reflection and second thoughts. Now he wondered: Why had the world become so vast? What wouldn't he give for a sight of John Phipps with little Joey tripping along? And who was this stroking his gnarled fingers in the dark, brushing his hair back from his eyes, drawing him to his feet, taking him back into the house where Mrs Covington had hot water bottles in the bed and a glass of hot rum ready?

More light in the room. Marine light on the ceiling, bouncing up from the harbour. With the light would come the sound of birds, if only Covington could hear them. A dim, deaf craving for the clarity of birdsong still touched him. His remaining senses were sharper: Sight. Smell. Taste. Touch. Let them have their run like the good sensations they were, he declared to himself, and let him be grateful for them and ignore the mill-race in his brain. His tea steamed on the window ledge, two sugar lumps on a plate beside it.

Theodora. What a blessing it was to have her in his life and be glad that he had never, not once, repudiated the claim Mrs FitzGerald made in her letter that she had a girl-child, and that Covington on all the evidence—strong nose, red hair, pale skin—was the father. From that very day and every birthday thereafter he sent money care of Merchant

Lumb in Buenos Ayres and was always concerned for her welfare. His mere oversight—his secret from his wife—had been in failing to tell Mrs Covington of Theodora's existence until Theodora was already on the water from California and coming to him. It had been shocking to Mrs Covington, shaking her faith in God's arrangements in her life—but good missus that she was she had recovered on meeting the young woman, oh yes indeed, and had not failed to love her. There seemed little doubt that Mac-Cracken was likewise smitten. With regard to the feelings of Theodora, however, Covington found himself struck dumb. He clasped his hands and recited a meditation long-tested in his life:

'This is like doing business in great waters. This is like being in the heart of the sea, and like going down to the bottoms of the mountains. Now it seems as if the earth, with its bars, were about us for ever. But let them that walk in darkness and have no light, trust in the name of the Lord, and stay upon their God. A-men.'

He dunked the sugar in his tea and when it was about to crumble popped it in his mouth and swallowed it down with a few hot sips.

'Make my porridge with cream,' he bellowed through to the kitchen, adding, 'An' I'll take more tea, me beauty!'—anticipating Mrs Covington coming through and buzzing around him so fondly as she always did when they made ready to travel. He unlaced his boots, getting himself geared for the day. Theodora wanted photographs before they left. She had arranged a studio, and it seemed like a good material boast, to have portraits. There she is my beautiful early-born daughter, secure in my heart, great with the spirit of life I gave her, carrier of my best blood, my sweet homing pigeon. But it worried him that the portraits might be her way of saying farewell. She had arrived on the *Betsy Blaine* out of San Francisco, and the second officer who was still in port kept sending her notes by messenger, she being open to persuasion as any woman might be—for give temptation an inch with 'em, he believed, and it became a bond. That officer—Covington had seen him—was quite, quite presentable. His ship hadn't hauled anchor yet, but still waited off Millers Point. The thought made Covington fume.

Theodora sat opposite him at the breakfast table. Confound my deafness. Confound my shyness with her, he boiled. The morning was chilly and she wore a long jacket with a fox-fur collar. It was tied at the throat with a brown string. Her hair was pulled back, pinned up, and strands fell down in natural curls. Her manners were straightforward, unaffected. Her small boots peeping out from under her chair were deep cherry-red, and her attaché-case, that she had placed near the door, far too small for a woman intent on inland travel. She sat very straight-backed, and in repose folded her hands in her lap. Compare her with the hoydens of Covington's later blood and she was a princess in life! His children shared red hair, freckles and fine aquiline noses (his own eagle-like nose, to be precise), and that was all.

He still could not ask whether she was consenting to coming with them today. What arrangements he had made he kept secret from them all. Wagons, beds, victuals: all the comforts of home (of ship, more like it) so they could travel down the land. Quite a caravan was in his mind. He peered from under his eyelids to read Theodora's lips and decipher her words. Her mother had taken her a merry dance through several continents. The last but one was India.

'More sugar ... father?'

Covington had the impression she begged his pardon. Then she repeated the question in mime.

'Yea, heartily, more sugar, pile it on.'

'He must have his sugar. It cheers him better than wine,' said Mrs Covington, so immensely considerate it stung.

He steered his wife aside.

'Is she comin'? Has she said yes?'

'Bless you. Ask her yourself.'

'Hmmm. Very well. Thank you.' What could you do with a wife who spoke your own thoughts but declined to pass on the very words of others?

A bar of morning sunlight warmed the room and Theodora stirred to its touch like a small ginger cat.

MacCracken? Where was he? Covington wanted to see the two of them together. What they looked like sitting at table. The ferrety MacCracken phizog against the somewhat rounder-faced, high-cheeked Theodora. Bone structure versus bone structure. Nose of one against the cheek of the other. Ribcage and breast contour. Their compatible strengths. Their weak points in opposition. Covington wanted to sense how they would go on from here—into the unknown of their lives. He wanted to know how they would take it, as a principle of life, if they had the news that the spirit was absent. *Stolen, torn out of the breathing world.* They would need to be strong, and when he thought of that he feared they might not shine. He feared nothing might shine. For himself no courage was wanting. But what was courage in this matter without hope?

There was no shortage of love in Covington towards his other children. The Australian Covingtons were a fine brood. They did the Queen's Empire credit, and even the youngest of them was now launched into the world without a second thought or a look back over her shoulder. They all loved their Pa in full return, and they knew him for a Pa of the very best—strong, boisterous, reliable, firm but kindly, always industrious, and sure to leave them generously bequeathed if he should drop off the perch. Yet perhaps a bit strange—more than so!—and therefore never to be quite understood. For they knew just as much of him as he wanted them to know. They knew about Darwin— Covington had named his second boy Charles Erasmus, after all. (Why? Because of damned superstition, that is why. Why else name a Congregationalist after a probable atheist?) There was no lack of pride in Covington's associations with quality, the Darwins, the FitzRoys, the Kings, even. But relating to deeper questions, as they touched on his children's knowledge of him, nothing more would come out of him than had already been given. They were good ridge timber, his progeny, and he was a stout trunk. They must never know about his querulous, leaf-withering side. They would find it stranger than a fencepost sprouting a head, or a cow talking English. The sons would take him for a milksop. But Theodora. But MacCracken. They prompted thoughts of a different quality altogether. They seemed to be his life and he hardly knew why, except the matter was pressing, and they had encompassing minds.

Was it to light a candle in his memory that he wanted this so? Was it to chisel his name on a granite slab placed vertically in the earth, and know its full meaning? Was it to tend his memorial days on a rise of land overlooking river flats where corn-tassels whispered, and white herons rose from an estuary and winged their way south along a line of misty surf? Was it to sit on the rough ground and tell him their news, whispering it to the grass? The strange wisdom of the future, which might turn on its head all that Covington found in his day? What grandchildren would come of them? What great-grandchildren? What strangers bearing his blood? Smaller and smaller proportions of his blood until infinitesimal but never quite gone—unless people were lost from the earth: someone to live the history of his blood in continuation of when it was first roused.

When the proposition was put to him, MacCracken believed he had never made such a swift and final decision in his life. Yes. He would go with the Covingtons to the country. This very morning on the coal-fired steam-packet that waited to chug them to Sydney Cove in the space of an hour. Then by pony trap and pack-horse as Mrs Covington sketched over breakfast, spooning him an egg and deferring to him like the dickens—miles over ridges and across tablelands and deep into the southern ranges until they arrived perched on the threshold of heaven, as he heard tell, overlooking both west and east. A good number of the cattle there were book possessions of MacCracken's already. It was where the land was, that Covington had spoken to him about some time before—when MacCracken had mentioned acreages and sheep as being worth more than gold, and Covington had nodded sagely, owning that he knew a man with four thousand acres and a fine house in a district with good soil and fair rainfall, who might be amenable to taking a new owner aboard. Was that man Covington himself? It mattered intensely to MacCracken if it was. For he had the infatuation that he would get Theodora thrown in. And so what if it meant abandoning every element of friendship, loyalty, and attachment—if it ripped to shreds MacCracken's reputation with his Afghans. This morning they felt like a faint dissembling set of feathers washed down the river of time, and MacCracken himself was a newer, stronger fellow. Reading Darwin was like getting a new pair of spectacles that sharpened MacCracken's view.

But where was the book? It was gone, and MacCracken had a faint hallucination of splashing the floor, missing his chamber-pot in the dark—dare saying that had been enough for Covington to interpret as an opinion, and take it away.

MacCracken's maid was called and given instructions. She was to hurry to 'Villa Rosa' and pack boots, cotton socks, weskits, regatta shirts, moleskin trousers, belts, cabbage tree hats and collar studs. Also his telescope, pistol, notebooks, tinder-box, pocket compass, negrohead tobacco, inks, pens, buffalo-hide medical bag and an assortment of dissecting knives for whatever might transpire. And his small shelf of poetry books, while she was at it, in consideration of romance, and his Galapagos sea-urchin spines on a saucer on the window ledge—the whole contents of that saucer, if she would, with its knick-knacks and treasures to interest a clever eye.

Mrs Covington made toast in front of the flames. It was how MacCracken liked it, dripping with butter and with a burr of charcoaled crumbs across the surface. She cut him a slice of pink ham and doled out mulligatawny chutney. But there was too much chewing involved. So she warmed stew in a pot of leftovers. This was better. It was the last dirty pot and when she scrubbed it and dried it and placed it in position over the fireplace she would have no chores left and be ready to leave 'Coral Sands' and lock the door.

While his stew warmed MacCracken took his cane and limped to the door of the cottage and sniffed the fresh air. Theodora had been in the house when he woke—he'd heard a melodious, good-humoured, accented voice, and had struggled from bed only to miss her. He shaded his eye against the sun. Was that her at the wharf, an intense impression of autumnal reddish-brown? Undoubtedly, and seeking to avoid him, too.

Covington appeared from the wharf pushing a four-wheeled baggage trolley.

'There you are, confound you for a creeping crippled man. Are you better?'

MacCracken demonstrated his agility, turning around and leaning inwards on his stick, making a full circle on the flagstones and looking (thought Covington with a certain satisfaction) like an animal caught in a trap without any realisation in the world that it was a trap indeed, and the bait was set, or rather would be if Theodora gave the nod.

MacCracken's jaw moved. He spoke or tried to speak. Covington came over and took his elbow, peered close and read his lips. Yes, there was movement along those bruises.

*I wish to slurry on the water.*

'What's that you say? A drink? You shall have better than *water*,' Covington boasted, putting his head through the doorway to direct his wife. 'Good God, man, you'll have breakfast. You shall have a soft-boiled egg and white bread, and whatever else you wish. Best China tea or Java coffee, and a mutton chop if your jaws can manage it, which I doubt they are able.'

So MacCracken kept what he'd really said to himself: *I love your daughter. I wish to marry your daughter.*

Then Covington was gone with a wave of his hat, and a loud shout: 'Be on the ten o'clock!'

MacCracken sucked shreds of meat between his newly chipped teeth. Physically matters were under control again. Ankle wobbly but holding, jaw manipulatory, and he was able to speak after a corner-of-the-mouth fashion.

'A little tea for you, doctor?'

'A great deal of tea, if you please.' It was needed to flush the opiates through. His vision was a little too bright. His mind was a carnival.

'An' how did you sleep?' Mrs Covington asked.

'Very well,' he nodded. Though his sinuses ached.

'Theodora saw you and said you tossed and turned. She stayed with Nurse Parkington last night.'

*It was her bed I took.* What a state of closeness he was in. No wonder his dreams had been wide-ranging and full of rapacity. Animals had figured prominently. Carnivorous animals well-armed, the shoulder pad of the boar, the hooked jaw of the male salmon. He woke in the thought that the shield might be as important for victory as the sword or spear—and obscurely gratified that the hearty Nurse Parkington was no special friend to Miss Georgina Ferris, and wouldn't take her side in a fit.

Of course if the Covingtons about-turned and announced that Theodora wouldn't be coming as hoped, then MacCracken might have to think twice. The thought of her staying in Sydney and being taken up, and wearing blue velvet at Government House balls and being subject to other men's lusts enraged MacCracken insensibly. He was full of zest this morning and cursed his injuries. He wanted to be wherever she was. He wanted to get another look at her, and Covington's deaf misunderstandings weren't far wrong—it was a thirst he had.

He thought if he could just glimpse her and work on that, he would be able to take in a little more, and so later look at her again, adjusting to the idea that he could work up to a conversation with her, and not feel incapable of speech twice over—once from the stiffness of his inferior maxillary, sore as perdition right up into the ramus, and the other from a most uncharacteristic fear that he wouldn't be able to charm her with his usual light tone, through shyness, bashfulness and lovestruck idiocy. His dreams of the night mocked him with their fancies. He would save his emotions and give her something—a trinket—that might serve to get them started. It did with all her sex as a rule.

Mrs Covington gathered herself into her compactness, brushed crumbs from her apron, straightened her skirts with a snap, and asked MacCracken what he thought of a man, full useless around the house and very demanding, who had once been better than a wife to another man.

'Mr Covington, you mean? Ha, was *that* what he was?'

'Oh, yes,' said the loving, affronted dame. 'Always doing the fetching and carrying. Answering every call, even cooking hot butter cakes and redding the poker last thing, to make a hot toddy. *I* never knew such a man. But then he told me,' Mrs Covington wiped away a tear, 'about his Theodora.'

'Ah, Theodora,' said MacCracken, just liking the name.

'We had a small scare with Mr Covington last night,' said Mrs Covington, getting on then and folding the napkins. 'At least Dorrie did. I keep calling Theodora Dorrie although you know *you* mustn't. Not in Mr Covington's hearing, that is.'

*How can I call her anything when I don't ever see her?* MacCracken had the peevish sensation of being abandoned that can afflict the person at the centre of everything.

'He is so very proud of her it would break your heart. I'm as wise to his manner as you are, Dr MacCracken, but Dorrie won't be told, and when it was well after dark she found him sitting on a rock. She said he was all steamed up from his conversation with you. *She* was a mite annoyed at your liberties, but I told her how needful my husband was, and of my trust in your part.'

MacCracken gravely nodded.

'They were not back for hours. I kept the lamps burning, as you know I do.'

MacCracken kept glancing through the windows, checking the door. He desperately wanted to be composed for Theodora if she came in. Perhaps she had left a memento behind. He felt pathetic about this. She might want to see him again. It was not inconceivable.

'Do you have any consolation?'

'I beg your pardon?' asked MacCracken.

'For my husband. Through what Mr Darwin has wrote. You know what hopes Mr Covington has for you, doctor, and heaven knows, I have gone up and down like a whirligig these past few years, as the time has drawn nearer to the book being done. It has been like coming towards a death in the man, I believe.'

'Well,' said MacCracken with difficulty, holding his jaw in his hand and finding the position made things easier, 'it is a wonderful piece of work. Mind I have only skimmed through—conditions were hardly conducive last night— um, *where* is your daughter, Mrs Covington?'

'My daughter?'

'Theodora.'

Mrs Covington coloured before replying. 'She has gone to the town with Covington, they have gone to see a photographer and we are to join them there. They are like each other. Always dashing around. Full of expectation and wonder.' She went to the window. 'Now I see that our bags are down at the jetty and I want you to take my arm. Because if we take it slow we'll have it easy together.'

Theodora had a knowledge of her father going back to the earliest, beginning moments of his life. Soon after she arrived in Sydney they walked for hours around the shore while he spoke of it. They sat in a tearoom while he spun it out. Sometimes she drifted off, nodding, losing the connection between a running band of urchins in an English field and the love of a young *vacciano* for her fickle, imprudent mother. Other times she sat upright with attention, and learned what she never knew— that Carreras, who was left, in his turn, by her mother, was one of Rosas's most assiduous Indian-killers, a disperser of bands of women and children to the most violent and sadistic ends. Yet a man with utter sentimentality over the loss of those savages, as she remembered him, a collector of their weavings and weapons and funereal stones.

Covington in turn learned of her: the excellent education she had from English nuns in a convent of Buenos Ayres. The unhappy marriage she made, that was annulled by Papal decree (she carried the dispensation with her). She told of the long journey she made to India with her mother, who had married, by then, an English general. It was where Theodora might have married someone she knew—having aided him in the cholera wards, and learned her Nightingale procedures under his instruction. India was where she finally decided that she would come to

Covington, take a chance in Australia, and answer his plea to be known.

Covington told her of his master. Of the greatness in him, and she had, indeed, heard his name.

'From where?'

She leaned across and pinched Covington's cheek.

'From mother.'

He told her that when Darwin was a boy he was locked in a long room with tall windows down either side, and ran with a stick smashing the windows, trying to get out, letting in the light. As for him, he had lately been feeling a darkness. There had been many letters from Darwin over the years. Covington showed her the bundle. 'See what I mean to him!'

Theodora read the letters over half an hour. There was a feeling through the tired friendliness and veiled condescension that Covington was envied—good health, no sickly or dead children, increasing property values, land, prospects for his offspring, progress in life. What if Darwin himself had emigrated? The thought was expressed. Australia might have been his choice. Now it was too late.

Theodora had tears in her eyes as she took Covington's hand. They sat in silence for a while.

To Covington, the simplicity of their moments was a great blessing, and no amount of thinking about how creation was *done* could take that away. With thoughts of the book flapping about in his head, dark as the wings of a bat, Covington basked in a light as clear as any star of his boyhood nights.

Covington told her about MacCracken. He laughed at that man. Theodora pulled a small face, and shook her head. 'Impossible.' Anyone who tried to make another shine in her eyes went about it the wrong way. The man she married would at first never seem the least dependent; yet would show her, in some fashion hardly known to himself, that he would cross the seas for her and walk on hot coals through adoration and love.

Covington wore a collarless cotton shirt and a black coat with the sleeves polished. Dark ridgelines curled around his right cheek, the gun-nuzzling side. A magnifying glass was needed to see them—those old, deep, connected stains, peppery as a Maorilander's tattoo—all from cradling the walnut stock of the small bore muzzle-loader that exploded in his face, lashing mustard-seed shot against him.

He clutched the book in his lap. The spine was showing. He was a man always obedient and with his sarcasm kept in check. Yet his silence bellowed:

'Be sure the volume can be seen.' He checked the inner pages, and pulled a sour face. 'It is the *second* printing!'

The title was plain. He wanted the photographer to be sure of it. *The Origin of Species*. How many thousand copies were made? If it was like Darwin's *Beagle's Voyage* there would be uncountable numbers of them, and in all the world's languages and libraries, too, and there would be no getting away from it anywhere.

Thoughts of himself went brawling like devils in his mind. He held the book tighter and wondered what to do to be part of a new fashion of believing. Shake a tree and spill the seed everywhere? Leap on a she-tiger? Go spread-eagled on the sea, trailing milt, disgorging phosphorescence between his legs?

Such monstrous pride in him and animal spirits.

He gave a warm laugh. How many issue must a man have to secure his place in creation? The answer was *enough*.

A light blinded him, a white potassium crunch of heat. Through the after-light came MacCracken wearing a floppy straw hat, and Theodora separately, her hair up in a radiant pile. They shall have the next one with him. Please. Stand here. You there. And 'Come on, darlin',' to the wife. It was a long ride he hoped for them all. The doughty missus. The doctor a son to him. The long-lost daughter malleable to his intentions. *If* she comes. And if she declines he will travel with MacCracken, and seek understanding, hoping she will follow later. But pray God she sees her way, or his heart might split.

'You must send one to your Don,' said MacCracken.

'He wants no photograph,' said Covington disparagingly.

'He asked for one. "With your next letter".'

'He never asks. He "wonders to trouble me". In that wonder is a demand. In that demand is a great opposition to being told no. Get your nose out of my letter box, MacCracken.'

'Give him an acknowledgement of your existence, old codger.'

'I send him barnacles packed in wooden crates,' said Covington, 'lined with tin, and he thanks me for remembering. I send him duck-skins and owl-skins. He praises the ship's captain who delivers them to his very front door without charge, never wondering if I have covered the freight generously, which I have indeed. I send him the foetus marsupials just the same.'

So he went on. It was much dry carping against a pretended grain, because here was the truth: under the complaints, the photograph, with Covington's new people, and with the book showing plain, was meant for Darwin all along. A fine expanded sense of Covington, have no doubt of it.

'What is the delay?' he shouted at the photographer, a shrivel-bearded man wearing an artist's smock. 'There is so little time. We have wagons to drive. Land to appraise.'

The cumbersome camera equipment stood ready once more. Covington fitted his wide round hat and stood with his stockwhip rampant. The photographer's assistant rumbled a scene to the back of them, a canvas frame on wheels, an oil-painted ridge of timber with sunbeams coming through and a team of thirty black bullocks straining to their limit.

'My friend Mr Earle was a painter,' declared Covington at large. 'He was a scabrous hound,' he added with great affection.

He angled his face for the camera and peered into the stalk-like shining lens. A good material boast. It cheered him spectacularly. A colonist of means with cattle and land, beauty and brawn in his blood, and a spate of thinkers, too. He made a vow. His bones would be read when he was gone—forehead, cheek, jaw—with more intelligence than they were in his lifetime; and his entire existence, like the prehistoric megatheriums of the Rio Colorado that the two of them found in a low cliff when they first started their life of shooting, would live on in importance. 'You don't know,' he muttered. 'You can't tell.'

'Are you talking to me?' said the doctor.

'Nay,' Covington answered, giving his friend a familiar look of disdain. 'Not you exactly ... ' Then shook himself and stared:

'I am ready.'

The photographer ignited the flare and the room filled with light.

Covington's hopes were present in the air like crystal. To get out of the room was like smashing a way through.

\*

Outside in the street the horses and dog-cart were ready, and Mrs Covington patted the cushions beside her, inviting MacCracken to ease himself up. Theodora, too, if it so pleased her—for the child was all indecision, and they were the three of them dependent on her decisions utterly, cravenly, bleakly, it felt. With Mr Covington riding there was room for the rest in the cart, Mrs Covington indicated, so very comfortable too, with a spring chassis, a taut canvas shade overhead, brass handgrips and rails, and strong ponies in the shafts fed on best stud-mash of Huddlestone's Livery Stables (in which Mr Covington happened to own a share, and likewise MacCracken). Theodora stood on the pavement bumped by passers-by. She looked distraught as she adjusted her bonnet. It was a shade of apricot, and very 'town'. They still weren't travelling clothes she was in. The others' trunks had gone ahead by dray, so there was plenty of space in the luggage-slide under the dog-cart. So would she please speak her mind?

'Mr Covington!' His wife tapped his shoulder with her whipstock, and he turned round to look at her.

'Yea?'

*Please go to her!* she mouthed.

But he wouldn't. Still couldn't. Just made a show of quieting his saddle-mare, putting his hand to its neck and running it smoothly down her flanks. It was in his heart to give love a push, but he felt the drawing back in himself. A quietening in him. Let the flower blossom or wither as it might. Even here in these messy, dusty streets. It was like turning his back on talk and listening for the stillness of God. Like pricking his ears—hearing the cry of birds—a wren in a thicket—where?

'My husband!' shrugged Mrs Covington, giving up on him.

MacCracken bowed to Theodora—she nodded briefly back—and they still had not spoken a word to each other, except for civilities, but addressed their every thought to the Covingtons.

Were they both smitten?

Neither thought the other so.

So both acted the same.

MacCracken climbed up beside Mrs Covington. She held the reins bunched in her hands, while the horses twitched away flies and stamped their feet, and asked, with their restlessness, if there was to be any get-going for them at all today.

Theodora bit her lower lip, looked to left and right, raised her toes and peered over the heads of pedestrians, jabbed her parasol-tip between a crack in the cobblestones and took it out again. Then she flicked her parasol open and shaded herself, because even though winter was coming the Sydney sun had bite. The parasol with its red fringes and shiny blue canopy was a fair landmark in the street. Anyone looking for her would notice it, and quite a few, who weren't looking for it, noticed it very appreciatively. MacCracken, owning a fair assessment of a lady's mind, reckoned to himself that Theodora had made arrangements with a beau to meet her outside the studio, to settle her mind on the question of this journey—say with a passionate declaration, or whatever.

A count to the sum of ten, or fifteen, no more, and Theodora nodded and snapped her brolly shut. All it took, then, was a moment of exasperated twirling and Theodora was done.

'I am coming,' she declared.

MacCracken would remember this about her and not get that flash of dismissal aimed at *him*. He could feel its quality and nothing had begun with them yet! He thought, if her green eyes settle on mine I'll snap, for I have no shield.

Theodora tucked her brolly under her arm and went to speak to Mr Covington. Practical business they had together; talk of horses; stables; trunks; hat boxes; and Covington smiled; kissed her cheek; Theodora laughed; took his hand; let it go; and in a moment Covington was up in the saddle, reaching down; and in a moment Theodora was up there behind him, sitting side-saddle, pillion-style, and laughing in high spirits.

'I know this laugh,' Covington was heard to say as he felt it vibrating through her arms.

'I wonder that he does,' muttered Mrs Covington. Theodora's mother, that Mrs FitzGerald, was a cat, *she* would have had a high, excited laugh that was full of the suprises of life, you could be sure.

'We are off to the stables to get Theodora a fine horse,' Covington announced, addressing the whole street it seemed and not just the occupants of the dog-cart in his loud pride. 'She will ride along with me, and you are to go to the shipping offices, m'dear, and fetch her trunks. Then we go to Liverpool town, clippety-clap, where there is a surprise waitin'.'

Thus dismissed, MacCracken and Mrs Covington looked at each other and raised their eyebrows. Mrs Covington gave a practised flick of greenhide, and the horses began wending their way along.

Early breakfast on their second day out. Misty sunlight through eucalyptus trees, the trunks a substantial yellow-gold and pale creamy blue, their bark curling down in copious scarves, the harsh cackling call of kookaburras marking their territorial districts in the trees, the twitching of wagtails in close, the drift of campfire smoke and the smell of new bread baking in the camp oven.

Mr Covington knelt at his devotions as he made a practice of doing each day, longer on Sundays, thanking God for his blessings and then getting up and brushing sticks and leaves from his knees.

Covington called them kooka*burrows*—he could faintly hear them, they were so raucous. He was excitable in boasting bush lore to the liveliest of the party (Theodora). It was a process that MacCracken found tedious when it got going, much of it rather too well known to him from his goldfields' tramps and other rural outings. It also denied him Theodora. But Theodora loved everything she learned. She was charmed by the new. How long would it last? he thought, and caught her eye. She held his in return, but only for a moment, and as if she had only a passing interest in him—not nearly so gripping as her fascination with a native beehive in a hollow limb, for example, when Covington reached up and jabbed it with a stick, causing the insects to fly out in a cloud. Or a lion-ant trap, when Covington led her by the hand, and they knelt in the dust examining it. Darwin had found one of these dusty cones, and said it was so like the European one that it steadied his faith in God. After their time in Sydney he had come back to the ship full of it: 'Now what would the disbeliever say to this? Would any two workmen ever hit on so beautiful, so simple and yet so artificial a contrivance? I cannot think so. The one hand has worked over the whole world.'

While Mrs Covington made breakfast over the fire, deftly managing pots, pans and cast-iron 'ovens', MacCracken propped himself on a trunk and delved deeper into Darwin's book, which he had grabbed back from Covington when they came into camp last night. It was a subject hardly begun between them. His concentration was only lightly on the ideas before him. He heard a dove calling in the paperbark trees, near the slow creek where they watered their horses, and thought about Darwin's use of pigeons as a breeding guide. He went back over a paragraph on the importance of large flocks for breed improvement. Then he jumped to the reason for lack of variety among donkey herds: they were kept by poor people, who lacked the leisure for seeking upward change. He peered at the draughthorses at the other end of the glade. Those carriers must be wealthy, he thought, to get them to such great perfection. Theodora came to the fire and helped Mrs Covington. She wore an open-collared shirt, a leather waistcoat, men's trousers, and an oval-crowned felt hat with the brim turned up. MacCracken was sick with love. Theodora had a manner of dealing with her father that MacCracken observed over the top of his pages, dropping his eyes whenever she glanced in his direction (but if she looked in his way he dropped his eyes indifferently). Hers was a warm, enthusiastic involvement with whatever Covington said, with whatever he proposed. 'Sugar in the fire sets it roaring ... bush honey is good for drawing boils ... use beeswax and mutton suet to water-proof your boots ... if you swallow a fishbone, mind, take a raw egg ... ' When his excitement touched Theodora she doubled it. If a puzzle was put—'How to boil water without a vessel?' 'Which way is south by the stars?'—she immediately set to working it out.

Their journey was to take longer than expected—sixteen days! There was to be no roughing it in wayside inns. There was a ladies' wagon and a men's wagon. The men's was loaded with farm supplies and Covington and MacCracken slept under it, with blankets wrapped round them and their heads 'on stones' as Covington boasted, though in fact they had small comfortable pillows stuffed with down, and their bed rolls, sheathed in canvas, were filled with bracken and most comfortable indeed.

But the ladies' wagon was more than delightful. It had two iron beds with mattresses, a sideboard with a mirror, washbasins, lamps, and a carpet on the floor. When they reached their destination the equipment would be put in the house. For now, it was a life afloat, for what was a wheeled wagon if not a boat on land?

Covington called MacCracken over to the camp fire, where Mrs Covington, sitting on a three-legged milking stool, doled out portions of breakfast. They stood around eating it. Drinking tea. Crunching on bacon. Tearing at mutton chops. MacCracken's jaw was much improved, but would words come out? Unlikely.

A pattern was set for the days ahead. There were two saddle-horses, two ponies for the dog-cart, and eight draughts, four to each wagon, driven by the taciturn carri-ers. These men were like sailors, giving a feeling of being friends of Covington's from some enterprise in the past when they had become sworn adherents of each other, after which they never needed to speak at length of anything again, just smoke their pipes, nod wisely in agreement, and play hands of cribbage when they got the chance, keeping score on stringybark shingles drilled with nail-holes, marked by twigs.

By nine they were on the road. Next morning they would make an earlier start. The air was chill, but on corners of the road, where dusty beams of sunlight poured through the trees, they felt a rising warmth. Covington rode with Theodora, MacCracken with Mrs Covington. Then they changed around. MacCracken and Covington sat in the ladies' wagon, comfortable on cushions, while Theodora and Mrs Covington took to the saddle-horses and cantered ahead, being found later walking them, Mrs Covington in a fine sweat but stimulated to the depths of her country heart; Theodora with a fine flush, her hair lying in sweaty strands on her lovely white neck.

'Your turn with Theodora, now,' boomed Covington, giving MacCracken a rather obvious nudge.

'I don't think so,' said the doctor. 'Really, I am most companionable.' He held up Darwin's book, 'I've a power of reading to do. Did you know, Covington,' he said, turning his back on Theodora and taking the man's arm, leaning on him a little as he limped through the dust in the wake of the creaking wagon, 'that beetles near a sea-coast, if they are non-flying, have less risk of being blown to sea, and so thrive better?'

'I did not know that,' grumped Covington. 'If you won't sit with Theodora you're a damned fool, man.'

'Then I am a damned fool,' said MacCracken.

'Aren't you enjoying our ride at all?'

'There is nothing in the whole wide world I would rather be doing,' said MacCracken, passionately sincere.

'I think you mean it,' said Covington, looking hard at his friend and feeling that MacCracken might have depths as yet unplumbable.

'Well, what am I to make of Darwin's book?' said Covington, then, lowering his voice a little, gripping MacCracken's arm, '*Is it proven?*'

But just then he saw Theodora reach up from the saddle and tear a strip of bark from a tree.

'Hrrumph,' said Covington in a pleased though critical fashion, and went to Theodora, helping her down from the horse and giving her a lesson in nature study. He fetched in his pocket for pill boxes. He had known from the start, he told her, not to stuff material in pockets and bring it back in a jumble. So he'd been an advantage to a naturalist often faced with toads flattened by cartwheels and nightjars devoured by ants and wrapped in leaf-litter. His Mr Darwin, he meant.

The days passed. The road climbed, wound, and emerged onto the frosty, sun-bright tablelands to the far south. Theodora and MacCracken knew what was to happen with them. When their backs were turned to each other they privately smiled, an inward, knowledgeable discovery—that this thing Covington wanted, as between two selected kinds, was what they wanted too. But wouldn't admit such a thing. Not to each other. All time was spread before them for the unfolding moment of love. 'Time *vaster*,' reflected MacCracken over their camp-fire evenings, with the cold stars bright overhead, 'than any of us have reason to know.'

They saw Covington at his prayers.

'Or that he knows, either.'

And was that what Darwin set out to prove—that limitless time made all things possible—once their voyage was over and they were settled at 36 Great Marlborough Street? Where sooty rain streaked the windows. Where daylight never showed, and it was full gloomy inside, but they had their many visitors—nitpickers, lady novelists with ear trumpets, geologists with horrible breath, and makers of infernal adding machines. They disturbed specimen trays behind Covington's back as if they had every right. They leafed through specimen lists and shifted labels from their rightful places, and said they 'hadn't meant to' when Covington roared them out, an apparition wearing a black dustcoat and a green eyeshade. They were the greats of English science and had no idea what Darwin was proposing. But Covington did. Just the two of them did: That species replaced each other in both time and place. That creatures occupying their various stations on earth had evolved from those that had gone before by a process of 'natural selection'. That species were living relatives of those that had gone before, just as they were of those still living nearby. That Noah, whose Ark was often compared with their cramped old vessel as a preserver of creation, was nothing but a bearded braggart whose tale was woven from hempen homespun.


'Seems I was some kind of South American fossil myself,' said Covington to MacCracken, coming back to their camp fire, 'and did not know what-for at all. I was smart as a carrot new-scraped and that was the best of me. I loved my scriptures and I lived my happiness to the full, but the more I was one thing, the more he was the other.'

# EPILOGUE

*On an Origin of Species*

∽

# December 1860

It was a headache that pierced to the marrow of Covington's brain. It came from staying up late with a smoking lamp and reading his Darwin, which he kept at his elbow with confused pride for a good few months after MacCracken and Theodora were married, and sailed for the States. It came from knuckling his eyes in the morning and going back over the passages again, and being rendered futile by understanding. It came from outbursts of anger—'It cannot be!'—'I will have the man flayed!'—'He *sneers* overmuch!'

It came from the effort of holding one book in his mind and remaining true to its texts while considering the rule of another. No amount of soothing from Mrs Covington's poultices helped, although when the pain receded, as it did for a time, he made a vow—he would have Christmas at Pambula, by the sea, and enjoy fresh corncobs from the crop he planted in spring.

So they came down the ridges from their upland acres with the cockatoo in a cage dangling from a dray, with Covington walking with the aid of a stick, demonstrating sure-footedness, don't you see—although secretly knowing that if he put too much weight on his left leg at the wrong moment he would go artichoke over turkey, as the saying was. He stumped along with a peg-leg motion and did his best. There was a numbness probing his senses like sea-urchin spines. When his headache returned it came writhing and shimmering like a sea-jelly in the shallows. When he took to his bed, at last, in a shaded room of 'Forest Oak', he felt easier in his mind, and rested his brain.

Then it came to Covington that his agitation was coming to an end, and he had better summon his spirits to him or there was no knowing what. For he was leaving home for ever, and doing it courageously, blindly—as such leavings were mostly done.

So it was in his memory that he ran to get first in line following John Phipps on a path leading south and westwards. They eased into a good rhythm along reed-fringed by-ways and cow-pads, soon climbing into rolling, chalky country where their feet were not constantly plunging into the mire. Phipps shouted questions, and his boys shouted back:

' "Is this the way to the Kingdom Come?" '

' "We are just in our way." '

' "How far is it thither?" '

' "Too far for any but those that shall get thither indeed." '

' "Is the way safe or dangerous?" '

' "Safe for those for whom it is to be safe, but transgressors shall fall therein." '

The catechism echoed a way of speech that no longer chimed with the rhythm of Covington's heart. It was the speech of his household and chapel left far behind. It was planted in his ear by the whispers of his mother. It was the speech of Mrs H with a delectable tickle affecting the pit of his stomach. It was the speech of the young man with gold buttons who pulled Covington up in the wall, leapt the stile, and spoke in his dreams. It was the speech spouted by Able Seaman John Phipps, a man whose meeting house was the open air and boundless horizons, whose congregation consisted of frogs in the wet grass, moles in the banks, spiders, snails, and the very birds tumbling through the air;

and also the fish in the sea if his reputation was anything to be reckoned with. And Covington had given it away for sixty pounds a year and a catechism in which was recited the families of creatures and the ladders by which small fitted to large, strange to familiar, until everything was sealed in a book that dared say it knew better than earnest old Bunyan's.

But still Phipps was present, and beckoned Joey Middleton up alongside him. One strode, the other trotted, the boy's breath going fast as a dragonfly's wings.

'Run, Joey, shall I catechise you?'

'Please do.'

'"Was there anything antecedent or before there was God?"'

'"No, for God is eternal. In six days the Lord made heaven and earth, the sea, and all that is in them."'

'"What do you think of the Bible?"'

'"It is the holy word of God."'

'Do you understand it?'

'Not a great deal.'

'What do you do with the places in it you don't understand?'

'I think God is wiser than I.'

'What about resurrection of the dead?'

'I believe they shall rise, the same that was buried. God has promised it.'

'And?'

'God is able to perform it.'

John Phipps turned back to the other boys—names forgotten, faces forgotten, all but spirits forgotten in time—and said to them in a loud voice, 'Joey has proved himself first.'

'But the last shall be first,' quipped the boy, who came last in the line.

'That,' said John Phipps stopping in his tracks, bringing the boy up to him quick smart, grabbing him by the collar, 'is because the best things are *for* the last, and do you know why?'

The boy spluttered he didn't. He expected it was humility.

'Because last must have his time to come. In eternity. "Therefore it is said of Dives. In thy lifetime thou receivedst thy good things, and likewise Lazarus evil things; but now he is comforted, and thou art tormented."'

Mr Covington was comforted. His paralysis creeping over him was a shadow covering the earth, while a star in the window burned white as diamonds.

When they reached their night's destination, miles beyond anywhere Covington had been in his life, past any last possible glimpse of Bedford's steeples, the boy Joey Middleton proved he was better than first. He ran ahead quick as a wisp to wait in the shelter of a lonely barn, with a fire of twigs going ready. He was more heart than sinew with his face gone blue from effort. Covington told him to stay by the fire, and gave him an apple and a sugar lump from his satchel. Joey gave Covington some portions of tobacco he had been saving. This was their friendship based in rivalry of attention that lasted past death.

A soft wind blew, sending hoops of rain across sodden green pastureland. Dark clumps of trees emerged and faded in the light. Rooks went down into their wood. A man with a gun could be seen in the distance walking in their direction. He was on the other side of a high stone wall that snaked across the valley, making a grim divider. After a time he stopped. Covington knew him; he was Darwin.

He called a black dog, and the dog went away with him. But after a time the dog was seen coming back. It was the sort of dog that was taught never to bark, only to slither forward on its belly like a puddle of tar, and then to spring out, yellow teeth and raking claws, and take a poacher by the arm.

It made John Phipps smile.

'I have no gun,' he said, reaching into his bag, 'but I have nets, lime, twigs, a small light—see?—and a bell.' He set out each of these things carefully.

'What about that dog?' wondered Covington.

'The dog stays close to his master,' said Phipps. 'It is what dogs are for. Otherwise they are all over the countryside. A dog will work itself to death if its master decrees. But later, when the master sleeps, the dog sleeps, and soundly too.'

Covington knew this for truth. A great sleep was his due portion.

'You thought you had a sailorman,' said one of the boys, 'but you have got yourself a fowler.'

'He is a poacher,' grinned Joey, squatting in friendly closeness in the circle. If John Phipps declared himself the Son of God at this moment they would hail the blasphemy as truth, because they worshipped him.

'A poacher in God's creation, I'll own to that,' said Phipps, tying a thread by holding one end in his teeth, and attaching the other to a pair of wires, then biting the thread clean off. 'This park was common pasture when I was a boy. What we had that was good then, was taken from us through acts of parliament. Whenever I pass I think it is in my power through the grace of God to take it back again.'

'Who is the man over there?' asked the boys.

'He is my brother.'

The night was mild and misty and still. Covington heard tassels of corn whispering on the river flats down from 'Forest Oak', and the cry of a plover and the distant sound of surf coming in from the Tasman Sea. The half moon rode above the haze of Pambula and Bedfordshire. Covington loved such nights, having experienced them in his independence, wandering the countryside and stemming his fear of hobgoblins and dark fiends just by the nature of who he was—a reasonable boy with no great harm in his breast. And for this reason no great harm came to him from any such encounters, and moonlit nights were the cradle of his being. He never imagined that to be reconciled would be to meet with himself this way, all times present in one. But it was why Mr Darwin was there, who had written in his book:

*If both are equally well fitted for their own places in nature, both probably will hold their own places and keep separate for almost any length of time.*

So the ancient species of believer still had a chance, it seemed.

Slowly the mist cleared and the valley was flooded with silvery light. Trees with their bare limbs and stately shapes shifted closer to the barn. They were sombre and deep. John Phipps yawned a prayer and said that just before moon-set, when shadows were long and clean, he would cross the valley and make his catch. It would be close to morning before he went. Now for sleep.

Covington lay back with his hands behind his head. The fire faded down. His lids grew heavy. Joey beside him shivered:

'It is cold 'tween the decks of a ship. Colder'n you could imagine, Cobby.'

'Then why are you mad to be going there?' whispered Covington back to him.

'It is hot in the Indies, that is why,' whispered Joey, and tucked himself into Covington's back, neat as a kitten. Then said, as if the thought was everything to him: 'I might see my Pa.'

Covington knew nothing more of that night except half-awake contentment each time he stirred and settled deeper in the bedding. It was the great night of his life. Others around him snored. He snored a good bit himself, giving Mrs Covington satisfaction and granting her respite from worry. When John Phipps crept away none of them knew he went. In the early morning Covington woke looking out from the broken doorway of the barn. The sun was not yet up. There was only an easing of the dark, the moon hanging like a tiny mirror back in a cleft of hills. Joey was up and had a fire going. In the first burst of cold sunlight John Phipps came strolling back, carrying a brace of plump birds that would keep them fed for the next sixty miles to Devonport and the ships of the Royal Navy.

The man he called his brother was over yonder. They ate hard cheese for breakfast and awaited his call. Soon Darwin came with his jacket buttoned to his neck and his eyes bright and excited. He gave scant attention to the birds with their feet tied. The two men embraced. There was a mood of withheld love between everyone present. A breeze stirred the cold, and John Phipps led them out to the shelter of a broad, bare-branched chestnut tree. Its trunk was like a wall, with roots making seats and footrests and arms for the weary. It was the old meeting place from a time when Congregationalists were driven from lawful worship. Many had gone to America where they had built a new Bedford. In the chilly sunlight overlooking the winter-green valley John Phipps spoke his words.

'Methinks here—under this tree—one may without much molestation be thinking what he is, whence he came, what he has done, and to what the King has called him.'

Covington trembled the words into his bones.

He saw Darwin on his knees, and there was no difference between prayer and pulling a worm from the grass. As for Mr Covington, he prayed in the old-fashioned way. It was the last of anything he knew.

# AUTHOR'S NOTE

*'My servant [Covington] is an odd sort of person; I do not very much like him; but he is, perhaps from his very oddity, very well adapted to all my purposes.'*
Charles Darwin in a letter to his sister,
from aboard HMS *Beagle*, 1834.

In basing *Mr Darwin's Shooter* on real people and actual events I have relied on many historical sources. Charles Darwin's archive is immense: he remains the most thoroughly documented scientific genius of the nineteenth century. Syms Covington's archive by comparison is tiny. It consists of a contested birth-date, a scrappy diary, a few watercolours, and scattered mentions in Darwin's letters and diaries. Yet Covington was at Darwin's side almost constantly from 1832 to 1839, during the voyage of the *Beagle* and for the two and a half crucial years following, when they lived in the same house and Darwin formulated his theory of natural selection in private notes. After Covington's emigration to Australia in 1839 they maintained a collecting relationship and a correspondence that ended with the arrival of *The Origin of Species* in Australia and, shortly afterwards, Covington's death ('of a paralysis') on 19 February 1861.

My principal Darwin sources have been: *The Origin of Species* (London, 1859); *The Red Notebook of Charles Darwin*, ed. Sandra Herbert (London, 1980); *The Correspondence of Charles Darwin, Volume I, 1821–1836* and *Volume II, 1837-1843*, eds Frederick Burkhardt and Sydney Smith (Cambridge, 1985 and 1986); and *Charles Darwin's Beagle Diary*, ed. Richard Darwin Keynes (Cambridge, 1988).

For Covington I used all of the above except the first. Also the following works: *Syms Covington of Pambula* by B. J. Ferguson (Merimbula, 1971, 1981); *Charles Darwin in Australia* by F. W. and J. M. Nicholas (Cambridge, 1989); and the unpublished diary of Syms Covington, held by the Mitchell Library, Sydney. I learned of Covington while reading *Darwin* by Adrian Desmond and James Moore (London, 1991). The most recent biography, *Charles Darwin: Voyaging* by Janet Browne (London, 1995), refers to Covington as 'the unacknowledged shadow behind [Darwin's] every triumph'. This is more my Covington than the diarist of somewhat undistinguished record he made himself in life.

Covington's belief, in this novel, speaks from *Pilgrim's Progress* by John Bunyan (London, 1678). His collecting would not have been possible, for me, without *The Naturalist's Guide for Collecting and Preserving Subjects of Natural History and Botany*, by William Swainson (London, 1822). Darwin almost certainly owned a copy of Swainson, and it is likely that it formed part of the library of the *Beagle*. For other aspects of bird collection I am indebted to the *Key to North American Birds* by Elliott Coules (Boston, 1903) and to Walter Bowles of the Australian Museum for directing me to it. My understanding of the importance of Covington's private collection of birds to Darwin's theory of natural selection is owed to two articles by Frank J. Sulloway. They can be found in the Spring and Fall 1982 issues of the *Journal of the History of Biology*: 'Darwin and His Finches: The Evolution of a Legend', and 'Darwin's Conversion: The *Beagle* Voyage and its Aftermath'.

I wish to thank Angela Marshall for the timely gift of a book; Rob Fenwick for the loan of one; Alan Gould for the same; Tony Milner for several; Trevor Shearston for Galapagos mementos; Lyndsay Brown for seafaring lore; Sue Fisher for invaluable manuscript readings and many clarifying conversations; Gerry Cassis for impromptu seminars in zoological lore, and for books, encouragement, and the wisdom of a naturalist; Linda Funnell for close editorial scrutiny; and last, but never least, Rose Creswell, my literary agent, for constant encouragement and enthusiastic guidance.